U.S. Marine Corps Close Combat

Plus

Desert Operations

Department of Defense

Pentagon
Publishing

DEPARTMENT OF THE NAVY
Headquarters United States Marine Corps
Washington, D.C. 20380-1775

18 February 1999

FOREWORD

1. PURPOSE

Today's Marines operate within a continuum of force where conflict may change from low intensity to high intensity over a matter of hours. Marines are also engaged in many military operations other than war, such as peacekeeping missions or noncombatant evacuation operations, where deadly force may not be authorized. During noncombative engagements, Marines must determine if a situation warrants applying deadly force. Sometimes Marines must decide in a matter of seconds because their lives or the lives of others depend on their actions. To make the right decision, Marines must understand both the lethal and nonlethal close combat techniques needed to handle the situation responsibly without escalating the violence unnecessarily. Marine Corps Reference Publication (MCRP) 3-02B, *Close Combat,* provides the tactics, techniques, and procedures of Marine Corps close combat. It also provides the doctrinal basis for the Marine Corps Close Combat Training Program (MCCCTP).

2. SCOPE

This publication guides individual Marines, unit leaders, and close combat instructors in the proper tactics, techniques, and procedures for close combat. MCRP 3-02B is not intended to replace supervision by appropriate unit leaders and close combat instruction by qualified instructors. Its role is to ensure standardization and execution of tactics, techniques, and procedures throughout the Marine Corps.

3. SUPERSESSION

MCRP 3-02B supersedes Fleet Marine Force Manual (FMFM) 0-7, *Close Combat,* dated 9 July 1993. There are significant differences between the two publications. MCRP 3-02B should be reviewed in its entirety.

4. WARNING

Techniques described in this publication can cause serious injury or death. Practical application in the training of these techniques will be conducted in strict accordance with approved Entry Level Close Combat, Close Combat Instructor (CCI), and Close Combat Instructor Trainer (CCIT) lesson plans. Where serious danger exists, the reac er is alerted by the following:

WARNING

5. CERTIFICATION

Reviewed and approved this date.

BY DIRECTION OF THE COMMANDANT OF THE MARINE CORPS

J. E. RHODES
Lieutenant General, U.S. Marine Corps
Commanding General
Marine Corps Combat Development Command

DISTRIBUTION: 14400006600

OVERVIEW OF CLOSE COMBAT

1. Purpose of Close Combat

Close combat is the physical confrontation between two or more opponents. It involves armed and unarmed and lethal and nonlethal fighting techniques that range from enforced compliance to deadly force. The purpose of close combat is to execute armed and unarmed techniques to produce both lethal and nonlethal results. Unarmed techniques include hand-to-hand combat and defense against hand-held weapons. Armed techniques include techniques applied with a rifle, bayonet, knife, baton, or any weapon of opportunity.

2. Continuum of Force

Marines will find themselves in both combative and noncombative situations. The threat level in these situations can rise and fall several times based on the actions of both Marines and the people involved. The escalation of force stops when one person complies with the demands imposed by another person. This range of actions is known as a continuum of force. Continuum of force is the concept that there is a wide range of possible actions, ranging from voice commands to application of deadly force, that may be used to gain and maintain control of a potentially dangerous situation (MCO 5500.6_, *Arming of Security and Law Enforcement [EE] Personnel and the Use of Force).* Continuum of force consists of five levels that correspond to the behavior of the people involved and the actions Marines use to handle the situation (see the table below). Close combat techniques are executed in levels three, four, and five.

Level One: Compliant(Cooperative)

The subject complies with verbal commands. Close combat techniques do not apply.

Level Two: Resistant (Passive)

The subject resists verbal commands but complies immediately to any contact controls. Close combat techniques do not apply.

Continuum of Force		
Level	**Description**	**Actions**
1	Compliant (Cooperative)	Verbal commands
2	Resistant (Passive)	Contact controls
3	Resistant (Active)	Compliance techniques
4	Assaultive (Bodily Harm)	Defensive tactics
5	Assaultive (Serious Bodily Harm/ Death)	Deadly force
Note: Shading indicates levels in which Marines use close combat techniques.		

Level Three: Resistant (Active)

The subject initially demonstrates physical resistance. Marines use compliance techniques to control the situation. Level three incorporates close combat techniques to physically force a subject to comply. Techniques include—

- Come-along holds.
- Soft-handed stunning blows.
- Pain compliance through joint manipulation and the use of pressure points.

Level Four: Assaultive (Bodily Harm)

The subject may physically attack Marines, but he does not use a weapon. Marines use defensive tactics to neutralize the threat. Defensive tactics include the following close combat techniques:

- Blocks.
- Strikes.
- Kicks.
- Enhanced pain compliance procedures.
- Nightstick blocks and blows.

Level Five: Assaultive (Serious Bodily Harm/Death)

The subject usually has a weapon and will either kill or seriously injure someone if he is not stopped immediately and brought under control. Typically, to control the subject, Marines apply deadly force through the use of a firearm, but they may also use armed and unarmed close combat techniques.

3. Marine Corps Tactical Concepts

Close combat techniques support the following key Marine Corps tactical concepts. The concepts are not standalone ideas but are to be combined to achieve an effect that is greater than their separate sum.

Achieving a Decision

Achieving a decision is important in close combat. An indecisive fight wastes energy and possibly Marines' lives. Whether the intent is to control an opponent through restraint or defend themselves in war, Marines must have a clear purpose before engaging in close combat and act decisively once engaged.

Gaining an Advantage

A basic principle of martial arts is to use the opponent' s strength and momentum against him to gain more leverage than one's own muscles alone can generate, thereby gaining an advantage. In close combat, Marines must exploit every advantage over an opponent to ensure a successful outcome. This can include employing various weapons and close combat techniques that will present a dilemma to an opponent. Achieving surprise can also greatly increase leverage. Marines try to achieve surprise through deception, stealth, and ambiguity.

Speed

Marines use speed to gain the initiative and advantage over the enemy. In close combat, the speed and violence of the attack against an opponent provides Marines with a distinct advantage. Marines must know and understand the basics of close combat so they can act instinctively with speed to execute close combat techniques.

Adapting

Close combat can be characterized by friction, uncertainty, disorder, and rapid change. Each situation is a unique combination of shifting factors that cannot be controlled with precision or certainty. For example, a crowd control mission may call for Marines to employ various techniques ranging from nonlethal restraint to more forceful applications. Marines who adapt quickly will have a significant advantage.

Exploiting Success

Typically, an enemy will not normally surrender simply because he was placed at a disadvantage. Marines cannot be satisfied with gaining an advantage in a close combat situation. They must exploit any advantage aggressively and ruthlessly until an opportunity arises to completely dominate the opponent. Marines must exploit success by using every advantage that can be gained.

(reverse blank)

Close Combat

Table of Contents

Page

Overview of Close Combat

Chapter 1. Fundamentals of Close Combat

1. Ranges of Close Combat 1-1
2. Weapons of the Body 1-2
3. Target Areas of the Body 1-3
4. Pressure Points of the Body 1-4
5. Basic Warrior Stance 1-6
6. Angles of Approach and Movement 1-7
7. Balance and Off-Balancing 1-8
8. Falls 1-9

Chapter 2. Lethal and Nonlethal Weapons Techniques

1. Bayonet Techniques 2-1
2. Nonlethal Rifle and Shotgun Retention Techniques 2-6
3. Nonlethal Handgun Retention Techniques 2-8
4. Firearm Disarmament Techniques 2-12

Chapter 3. Hand-Held Weapons

1. Fundamentals of Knife Fighting 3-1
2. Knife Fighting Techniques 3-3
3. Weapons of Opportunity 3-6
4. Fundamentals of Combative Stick 3-6
5. Combative Stick Techniques 3-7
6. Blocking Techniques 3-9
7. Unarmed Against Hand-Held Weapons 3-11
8. Counters to Hand-Held Weapon Attacks 3-13

Chapter 4. Strikes

1. Principles of Punches 4-1
2. Punches 4-2
3. Strikes with the Upper Body 4-4
4. Strikes with the Lower Body 4-10
5. Counters to Strikes 4-12

Page

Chapter 5. Throws

1. Turning Throw 5-1
2. Hip Throw 5-2
3. Leg Sweep 5-3

Chapter 6. Chokes and Holds

1. Types of Chokes 6-1
2. Chokes 6-1
3. Counters to Chokes and Holds 6-5

Chapter 7. Ground Fighting

1. Offensive Ground Fighting 7-1
2. Defensive Ground Fighting 7-3
3. Ground Fighting Chokes 7-5

Chapter 8. Nonlethal Techniques

1. Unarmed Restraints and Manipulation 8-1
2. Nonlethal Baton 8-9

Appendices

A. Pugil Stick Training A-1
B. Safety Precautions During Training B-1

CHAPTER 1

FUNDAMENTALS OF CLOSE COMBAT

This chapter describes all techniques for a right-handed person. However, all techniques can be executed from eittwr side.

The Marine is depicted in camouflage utilities. The opponent is depicted without camouflage.

The fundamentals of close combat include ranges, weapons of the body, target areas of the body, and pressure points of the body. These fundamentals form the basis for all close combat techniques. They provide Marines with a common framework regardless of the type of confrontation or the techniques used. If Marines apply these fundamentals properly in a close combat situation, they may save their lives or the lives of fellow Marines.

1. Ranges of Close Combat

Close combat engagements occur within three ranges: long range, midrange, and close range. During any engagement, these ranges may blur together or may rapidly transition from one to another until either the opponent is defeated or the conflict is resolved.

Long Range

During long range engagements, combatants engage each other with rifles, bayonets, sticks, or entrenching tools. See figure below.

Midrange

During midrange engagements, combatants engage each other with knives, punches, or kicks.

Close Range

During close range engagements, combatants grab each other. Close range engagements also involve elbow strikes, knee strikes, and grappling.

2. Weapons of the Body

Hands and Arms

The hands, forearms, and elbows are the arm's individual weapons. The hands consist of several areas that can be used as weapons: fists, edges of hands, palms, and fingers.

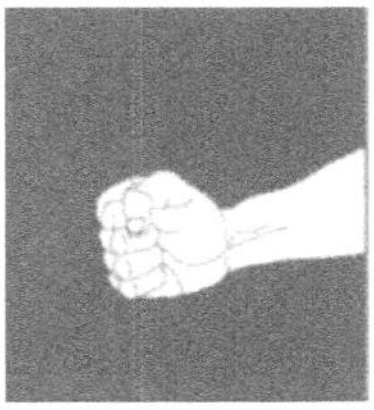

Fists. To minimize injury to the fists, Marines use their fists as weapons to target soft tissue areas such as the throat. The fists' striking surfaces are the first two knuckles of the hands or the meaty portions of the hands below the little fingers.

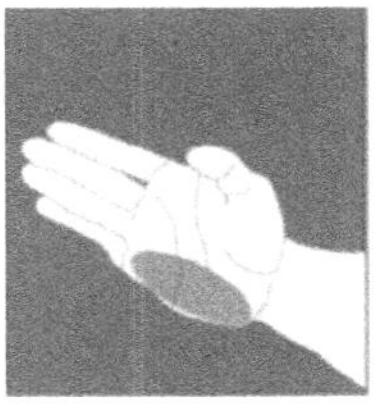

Edge of Hand. Marines use the edge of the hand (knife edge) as a weapon. Marines use the edge of the hand to strike soft tissue areas.

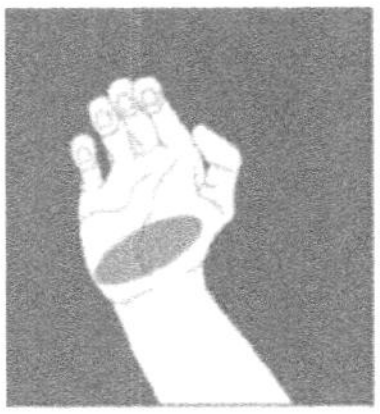

Palms. Because of the palm's padding, Marines use the heels of the palms to strike, parry, and/or block.

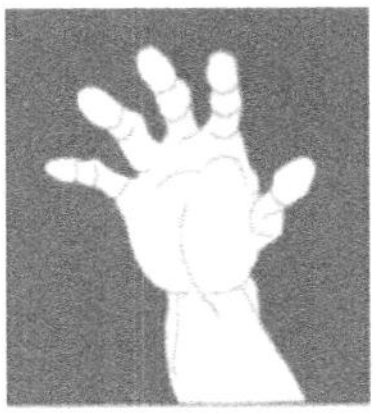

Fingers. Marines use the fingers to gouge, rip, and tear soft tissue areas (e.g., eyes, throat, groin).

Forearms. Marines use the forearms as a defensive tool to deflect or block attacks. Forearms can also be used as striking weapons to damage or break an opponent's joints and limbs. Marines sustain less self-injury when strikes are conducted with the forearms than when strikes are conducted with fists and fingers.

Elbows. Marines use the elbows as striking weapons. Because of the short distance needed to generate power, elbows are excellent weapons for striking during the close range of close combat.

Legs

The legs are more powerful than any other weapon of the body, and they are less prone to injury when striking. The feet are protected by boots and are the preferred choice for striking.

Feet. Marines use the balls of the feet, the insteps, and the toes to kick an opponent. Marines use the cutting edge of the heels and the heels to stomp on an opponent. Marines must be wearing boots when striking with the toes.

Knees. Like elbows, knees are excellent weapons in the close range of close combat. Knee strikes are most effective while fighting close to an opponent where kicks arc impractical. The opponent's groin area is an ideal target for the knee strike if he is standing upright. Knee strikes can deliver a devastating secondary attack to an opponent's face following an initial attack that caused him to bend at the waist.

3. Target Areas of the Body

During close combat, Marines strive to attack the accessible target areas of an opponent's body. The readily accessible areas will vary with each situation and throughout the engagement. The target areas are divided into five major groups: head, neck, torso, groin, and extremities. The figure below illustrates target areas of the body.

Head

The vulnerable regions of the head are the eyes, temple, nose, ears, and jaw. Massive damage to the head kills an opponent.

Eyes. The eyes are excellent targets because they are soft tissue areas that arc not protected by bone or muscle. Attacks to this area may cause the opponent to protect the area with his hands, allowing Marines to execute a secondary attack to other

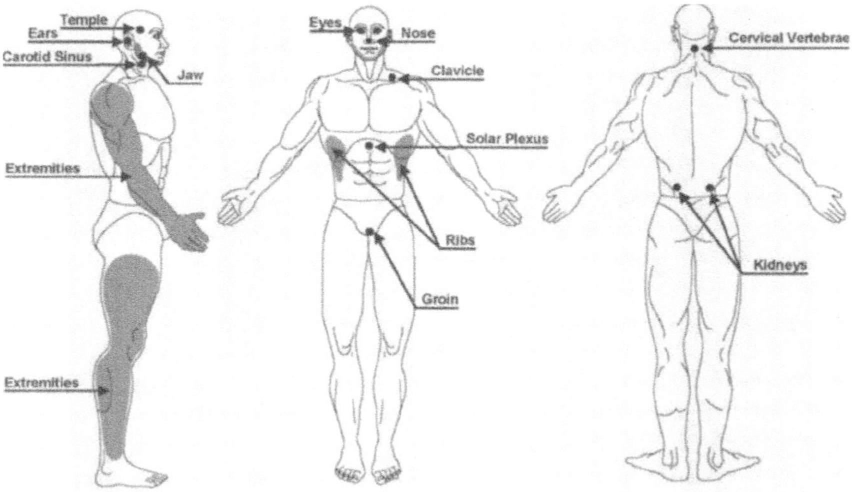

target areas while the opponent uses his hands to protect his eyes.

Temple. The temple is one of the most fragile areas of the head. Powerful strikes to the opponent's temple cause permanent damage and death.

Nose. The nose is very sensitive and easily broken. An attack to this area causes involuntary watering and closing of the opponent's eyes, rendering him vulnerable to secondary attacks. However, through training, individuals can condition themselves to withstand attacks to the nose. Therefore, any attack to the nose must be powerfully delivered.

Ears. Attacks to the ears may cause the eardrum to rupture. But this may not stop or even distract an opponent unless Marines powerfully deliver the strike.

Jaw. The jaw region, when struck forcefully, renders the opponent unconscious. Strikes to the jaw cause painful injuries to the teeth and surrounding tissues (e.g., lips, tongue), but the risk of self-injury is great unless Marines deliver strikes with a hard object such as a helmet, rifle butt, or boot heel.

Neck

The front of the neck, or throat area, is a soft tissue area that is not covered by natural protection. Damage to this region causes the opponent's trachea to swell, closing his airway, which can lead to death.

Carctid Sinus. The carotid sinus is located on both sides of the neck just below the jaw. Strikes to the carotid sinus restrict blood flow to the brain, causing loss of consciousness or death.

Cervical Vertebrae. The cervical vertebrae on the back of the neck, from the base of the skull to the top of the shoulders, contains the spinal cord, which is the nervous system's link to the brain. The weight of the head and the lack of large muscle mass allow damage to the cervical vertebrae and spinal cord. Excessive damage to this area causes pain, paralysis, or death.

Torso

Clavicle. The opponent's clavicle (or collar bone) can be easily fractured, causing immobilization of the arm.

Solar Plexus. Attacks to the opponent's solar plexus or center of the chest can knock the breath out of him and immobilize him.

Ribs. Damage to the opponent's ribs immobilizes him. It may also cause internal trauma.

Kidneys. Powerful attacks to the opponent's kidneys cause immobilization, permanent damage, or death.

Groin

The groin area is another soft tissue area not covered by natural protection. Any damage to this area causes the opponent to involuntarily protect his injured area, usually with his hands or legs. In male opponents, the scrotum is the main target since even a near miss causes severe pain, contraction of the lower abdominal muscles, deterioration of his stance, and possible internal trauma.

Extremities

Rarely will an attack to the opponent's extremities (arms and legs) cause death, but they are important target areas in close combat. Damage to an opponent's joints causes immobilization.

4. Pressure Points of the Body

There are nerves in the human body that, when pressure is applied or when they are struck, allow Marines to control a subject through pain compliance. Marines use pressure points to control an opponent when deadly force is not authorized. They also use pressure points to soften or distract an opponent so a lethal or nonlethal technique can be employed. The figure on page 1-5 illustrates the body's pressure points. Marines execute attacks to pressure points by—

: Rapidly kicking or striking pressure points.
: Slowly applying steady pressure to pressure points.

Infraorbital Nerve

The infraorbital nerve is just below the nose. Marines apply pressure to this nerve with an index finger to control the opponent.

Mastoid Process

The mastoid process is behind the base of the ear and beneath the edge of the jaw. Marines apply inward and upward pressure to this pressure point with the fingers to distract and control the opponent.

Jugular Notch

The jugular notch is at the base of the neck in the notch formed at the center of the clavicle. Marines apply pressure in a quick, stabbing motion with the index finger. Strikes to the jugular notch cause serious damage.

Brachial Plexus (Tie In)

The brachial plexus (tie in) is on the front of the shoulder at the joint. Strikes and pressure applied with the hand are effective on this nerve.

Radial Nerves

Radial nerves are on the inside of the forearms along the radius bones. Strikes and pressure applied with the hand to the radial nerve serve as a softening technique.

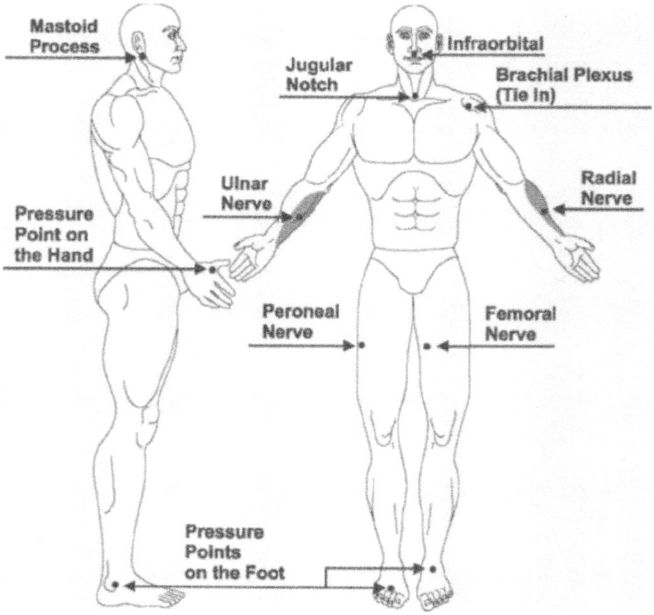

Ulnar Nerve

Ulnar nerves are on the outside of the forearms along the ulnar bones. Strikes and pressure applied with the hand to the ulnar nerve serve as a softening technique.

Pressure Point on the Hand

The hands contain a pressure point on the webbing between the thumbs and index fingers where the two bones of the fingers meet.

To force an opponent to soften or release his grip, Marines apply pressure with their index fingers to this pressure point or strike this pressure point with their fists.

Femoral Nerves

Femoral nerves are on the inside of the thighs along the femur bones. Strikes to the femoral nerve serve as a softening technique.

Peroneal Nerves

Peroneal nerves are on the outside of the thighs along the femur bones. Strikes to the peroneal nerve serve as a softening technique.

Pressure Points on the Feet

There are pressure points on the feet that, when pressure is applied or when they are struck, serve to soften or distract the opponent. Marines apply pressure with the toe, edge, or heel of their boots to the following points:

: The notch below the ball of the ankle.
: The top center of the foot, above the toes.
: The top of the foot where the leg and foot meet.

5. Basic Warrior Stance

Achieving a solid stance is critical to stability and movement throughout any close combat confrontation. The basic warrior stance provides the foundation for stability and movement that is needed to execute close combat techniques. To execute the basic warrior stance, Marines put their feet apart, hands up, elbows in, and chin down.

Feet Apart

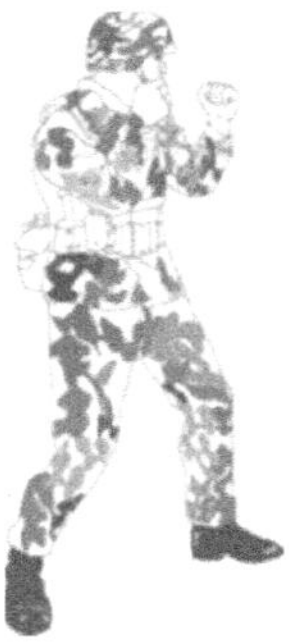

Place feet shoulder-width apart.

Keep the head forward and eyes on the opponent, take a half step forward with the left foot, and pivot on the heels so the hips and shoulders are at approximately a 45-degree angle to the right.

Distribute body weight evenly on both legs. Bend the knees slightly.

Hands Up

Curl the fingers naturally into the palm of the hand. Position the thumb across the index and middle fingers. Do not clench the fists. Clenching the fists increases muscular tension in the forearms and decreases speed and reaction time.

Bring the hands up to the face at chin level, with the palms facing each other. Hold the fists up high enough to protect the head, but not so high that they block the field of vision. Ensure continuous eye contact with the opponent.

Elbows In

Tuck the elbows in close to protect the body.

Chin Down

Tuck the chin down to take advantage of the natural protection provided by the shoulders.

6. Angles of Approach and Movement

Marines use movement to control a confrontation and to retain a tactical advantage. Movement increases power and maximizes momentum. By moving around the opponent, Marines gain access to different target areas of the opponent's body. Movement allows Marines to use different weapons of the body and different close combat techniques to attack specific target areas.

Angles of Approach

Marines move anywhere within a 360-degree circle around the opponent to gain a tactical advantage. This circle provides access to different target areas of the opponent's body.

When facing an opponent, Marines move in a 45-degree angle to either side of the opponent. Moving at a 45-degree angle avoids an opponent's strike and puts Marines in the best position to attack the opponent. Marines should avoid being directly in front of an opponent during a confrontation. If a Marine is directly in front of an opponent, the opponent can rely on his forward momentum and linear power to seize the tactical advantage.

Movement

Marines must know how to move in all directions while maintaining the basic warrior stance. During any movement, the legs or feet should not be crossed. Once a movement is completed, the basic warrior stance should be resumed. Maintaining the basic warrior stance protects Marines and puts them in the proper position to launch an attack against an opponent.

Note: Before body movement begins. Marines turn their heads quickly to the new direction. The faster the head turns, the faster the body moves, and the quicker Marines attain visual contact with the opponent.

Forward to the Left. To move forward to the left from the basic warrior stance, Marines-

: Move the left foot forward at a 45-degree angle from the body (approximately 12 to 15 inches), keeping the toe pointed toward the opponent.

Bring the right foot behind the left foot as soon as the left foot is in place. This returns Marines to the basic warrior stance.

Forward to the Right. To move forward to the right from the basic warrior stance, Marines-

Move the right foot forward at a 45-degree angle from the body (approximately 12 to 15 inches).

Bring the left foot, toe pointing toward the opponent, in front of the right foot as soon as the right foot is in place. This returns Marines to the basic warrior stance.

Backward to the Left. To move backward to the left from the basic warrior stance, Marines execute the forward movement in reverse. Marines-

Move the left foot backward at a 45-degree angle from the body (approximately 12 to 15 inches), keeping the toe pointed toward the opponent.

Bring the right foot behind the left foot as soon as the left foot is in place. This returns Marines to the basic warrior stance.

Backward to the Right. To move backward to the right from the basic warrior stance, Marines execute the forward movement in reverse. Marines-

Move the right foot backward at a 45-degree angle from the body (approximately 12 to 15 inches).

Bring the left foot, toe pointing toward the opponent, in front of the right foot as soon as the right foot is in place. This returns Marines to the basic warrior stance.

7. Balance and Off-Balancing

Balance

In any close combat situation, Marines must strive to maintain balance. The last place to be in a close combat situation is on the ground. Marines must maintain a strong base and a low center of balance, their feet must be a shoulder-width apart, and they must stay on their toes to enable quick movement.

Off-balancing

Marines use off-balancing techniques to control an opponent. These techniques are used to throw an opponent to the ground while Marines remain standing, or they are used to put Marines in a position for an offensive attack.

Off-balancing techniques use the opponent's momentum to move or throw him. For example, if the opponent is charging a Marine, the Marine pulls the opponent to drive him to the ground. Likewise, if the opponent is pulling a Marine, the Marine pushes the opponent to drive him to the ground.

Off-balancing techniques also rely on the power generated by the opponent. For example, during combat a Marine may be tired or outnumbered. Depending on the generated energy and momentum of the opponent, the Marine employs off-balancing techniques with very little effort and still provides effective results.

Because off-balancing techniques rely on the momentum and power generated by the opponent, these techniques are particularly effective for Marines who may be outsized by their opponent or lack their opponent's strength.

Angles of Off-balancing

There are eight angles or directions in which an opponent can be off-balanced: forward, rear, right, left, forward right, forward left, rear right, and rear left.

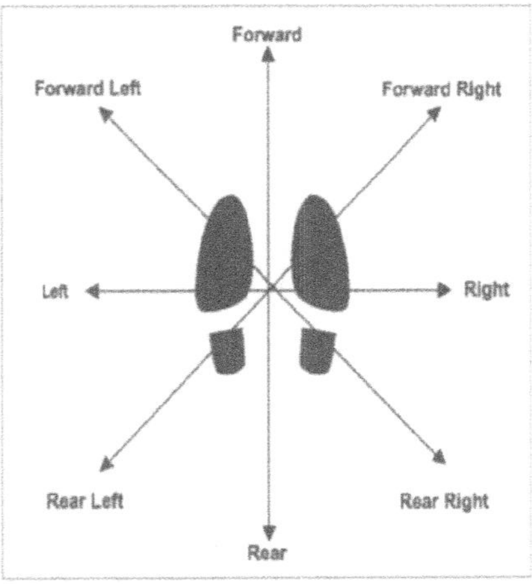

*Note: The angles correspond to the Marine's perspective, not the opponent's. Forward, rear, right, and left are straight angles. Forward right, forward left, rear right, and rear left are considered quadrants that are at a **45**-degree angle in either direction to the front or rear.*

Off-balancing Techniques

Marines off-balance an opponent by pushing, pulling, or bumping the opponent with their hands, arms, or bodies.

To pull, Marines grab an opponent with their hands and drive him forcefully to one of the rear quadrants or to the right or left.

To push, Marines grab the opponent with their hands and drive him forcefully into one of the front quadrants or to the right or left.

Marines execute bumping in the same manner as pushing, but use their shoulders, hips, and legs instead of their hands to off-balance the opponent.

8. Falls

Marines may lose their balance or be thrown to the ground during encounters with an opponent. Marines use falling techniques to absorb the impact of a fall and to quickly return to their feet following an opponent's attack.

Whether falling or being thrown by an opponent, Marines strive to reduce the force of the impact, to prevent serious personal injury, and to increase the chances of survival. Falling techniques use the body's large muscles (back, thighs, buttocks) to protect vital organs and bones from injury and immobilization.

Front Fall

Marines execute a front fall to break a fall on the front. To execute the front fall, Marines-

Bend the elbows and place the palms facing out in a position to spread and absorb the impact of the fall.

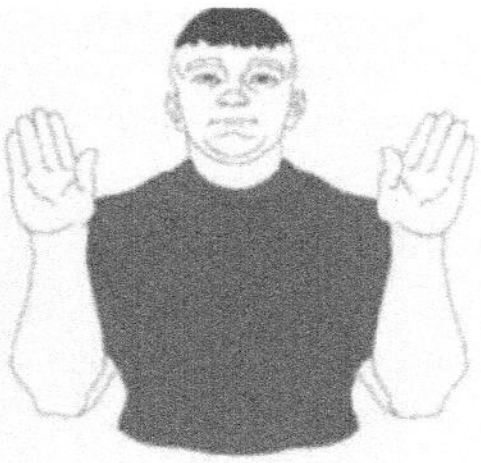

Fall forward, breaking the fall with the forearms and palms. The forearms and hands, down to the fingertips, should strike the ground simultaneously.

Offer resistance with the forearms and hands to keep the head raised off the ground.

Side Fall

Marines execute a side fall to break a fall on the side. To execute the side fall, Marines-

Bring the right arm across the body so the hand is next to the left shoulder with the palm facing inboard.

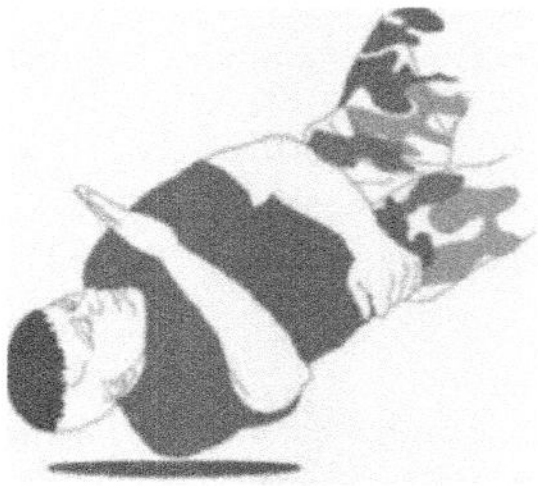

Fall to the side, breaking the fall with the right arm by slapping the ground and making contact from the shoulder or forearm down to the hand. At the same time, tuck the chin and keep the head raised off the ground. The chin should be tucked to the chest at all times to prevent whiplash.

Stretch the right leg out to make contact with the ground and to distribute and absorb the impact. Bend the left leg, allowing the foot to make contact with the ground.

Back Fall

Marines execute a back fall to break the fall when being thrown or falling backward. To execute the backfall, Marines-

Cross the hands in front of the chest and tuck the chin.

Fall backward and slap the ground with the forearms and hands to absorb the impact of the fall and keep the head off the ground.

Forward Shoulder Roll

Marines use the forward shoulder roll to break a fall from an opponent's attack and to use the momentum of the fall to get back on their feet quickly. Ideally, Marines execute the forward shoulder roll to a standing position so they can continue fighting. To execute the forward shoulder roll, Marines-

Contact the ground with the back of the right forearm and upper arm. Tuck the chin into the chest.

Roll onto the right shoulder, rolling diagonally across the back to land on the left hip.

Slap the ground with the left arm, absorbing the impact from the shoulder to the hand, palm down.

Keep the left leg straight to absorb as much of the impact as possible. The right leg is bent and the foot hits flat on the ground.

Bend the left leg upon impact to push off with the left knee and leg to a squatting and then a standing position. Forward momentum should carry the Marine to a standing position.

CHAPTER 2

LETHAL AND NONLETHAL WEAPONS TECHNIQUES

This chapter describes all techniques for a right-handed person. However, all techniques can be executed from either side.

In drawings, the Marine is depicted in woodland camouflage utilities; the opponent is depicted without camouflage. In photographs, the Marine is depicted in woodland camouflage utilities; the opponent is depicted in desert camouflage utilities.

1. Bayonet Techniques

WARNING

During training, Marines must have bayonets sheathed. Marines use bayonet dummies to practice bayonet techniques. When practicing offensive and defensive bayonet techniques student-on-student, Marines use pugil sticks.

All Marines armed with a rifle carry a bayonet. The bayonet is an effective weapon if Marines are properly trained in offensive and defensive bayonet techniques. An offensive attack, such as a thrust, is a devastating attack that can quickly end a fight. Defensive techniques, such as the block and parry, can deter the opponent's attack and allow Marines to regain the initiative. Through proper training, Marines develop the courage and confidence required to effectively use a bayonet to protect themselves and destroy the enemy. In situations where friendly and enemy troops are closely mingled and rifle fire and grenades are impractical, the bayonet becomes the weapon of choice.

Holding the Rifle

To execute bayonet techniques, Marines hold the rifle in a modified basic warrior stance. All movement begins and ends with the basic warrior stance. To hold the rifle, Marines—

Use an overhanded grasp to grab the small of the rifle's stock. Use an underhanded grasp to grab the handguards of the rifle.

Lock the buttstock of the rifle against the hip with the right forearm.

Orient the blade end of the rifle toward the opponent.

Offensive Bayonet Techniques

Straight Thrust. Marines use the straight thrust to disable or kill an opponent. It is the most deadly offensive technique because it causes the most trauma to an opponent. Target areas are the opponent' s throat, groin, or face. The opponent's chest and stomach are also excellent target areas if not protected by body armor or combat equipment. To execute the straight thrust, Marines—

Lift the left leg and lunge forward off the ball of the right foot while thrusting the blade end of the weapon forward, directly toward the opponent.

Retract the weapon and return to the basic warrior stance.

Slash. Marines use the slash to kill an opponent or to create an opening in his defense. The target area is the opponent's neck. To execute the slash, Marines—

Extend the left hand back toward the left shoulder.

Thrust the left hand forward and swing it to the right, bring the right hand back toward the hip, and turn the cutting edge of the blade toward the opponent's neck. The movement is a slashing motion so the blade cuts across the opponent's neck.

Horizontal Buttstroke. Marines use the horizontal buttstroke to weaken an opponent's defenses, to cause serious injury, or to set him up for a killing blow. Target areas are the opponent's head, neck, and legs. To execute the horizontal buttstroke, Marines—

Step forward with the right foot and drive the right hand forward. Rotate the hips and shoulders into the strike. Move the left hand back toward the left shoulder.

Strike the opponent with the butt of the weapon.

Vertical Buttstroke. Marines use the vertical buttstroke to weaken an opponent's defenses, to cause serious injury, or to set him up for a killing blow. Target areas are the opponent's groin and face. To execute the vertical buttstroke, Marines—

Step forward with the right foot and drive the right hand straight up.

Pull the left hand back over the left shoulder.

Strike the opponent with the butt of the weapon.

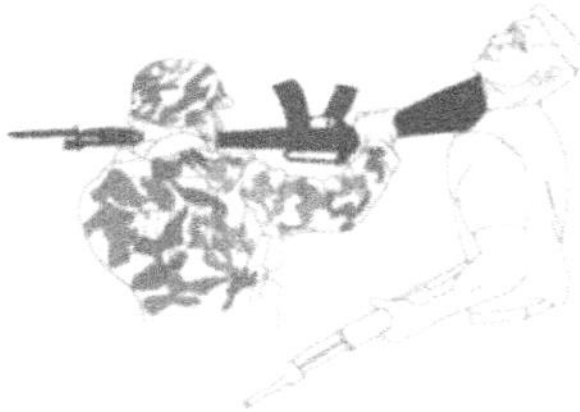

Smash. Marines use the smash as a follow-up technique to the horizontal or vertical buttstroke, particularly if they missed the target. The target area is the opponent's head. To execute the smash following a buttstroke, Marines-

Step forward with the right foot and place the blade end of the weapon over the left shoulder and elevate the right elbow above the shoulders.

Drive the arms straight forward, striking the opponent with the butt of the weapon.

Defensive Bayonet Techniques

Parry. Marines use a parry as a defensive technique to redirect or deflect an attack. A parry is a slight redirection of a linear attack by an opponent; e.g., a straight thrust or a smash. To execute the parry, Marines-

Use the bayonet end of the rifle to redirect the barrel or bayonet of the opponent's weapon.

Lock the weapon against the hip with the right forearm.

Rotate to the right or left, moving the bayonet end of the rifle to parry the opponent's attack. Rotation is generated from the hips.

Redirect or guide the opponent's weapon away from the body by exerting pressure against the opponent' s weapon.

High Block. Marines execute a high block against a vertical attack coming from high to low. To execute the high block, Marines—

Thrust the arms up forcefully at approximately a 45-degree angle from the body. The weapon should be over the top of the head and parallel to the ground. The elbows are bent, but there is enough muscular tension in the arms to absorb the impact and deter the attack.

Low Block. Marines execute the low block

against a vertical attack coming from low to high. To execute the low block, Marines—

Thrust the arms down forcefully at approximately a 45-dcgrcc angle from the body. The weapon

should be at or below the waist and parallel to the ground. The elbows are bent, but there is enough muscular tension in the arms to absorb the impact and deter the attack.

Left and Right Block. Marines execute a left or right block against a horizontal buttstroke or a slash. To execute the left or right block, Marines-

Thrust the arms forcefully to the right or left, holding the rifle vertically in the direction of the attack. The elbows are bent, but there is enough muscular tension in the arms to absorb the impact and deter the attack.

Counter Action Following the Block. After deflecting an opponent's attack with a block, Marines counter with a slash or a horizontal buttstroke to regain the initiative. However, the objective in any bayonet fight is to thrust forward with the blade end of the weapon to immediately end the fight.

Group Strategy

On occasion, Marines may engage an opponent as a member of a group or numerous opponents by one's self or as a member of a group. By combining bayonet fighting movements and simple strategies, Marines can effectively overcome their opponent or opponents.

Offensive Strategy: Two Against One. If two bayonet fighters engage one opponent, the fighters advance together.

Fighter 1 engages the opponent while fighter 2 swiftly and aggressively attacks the opponent's exposed flank and destroys the opponent.

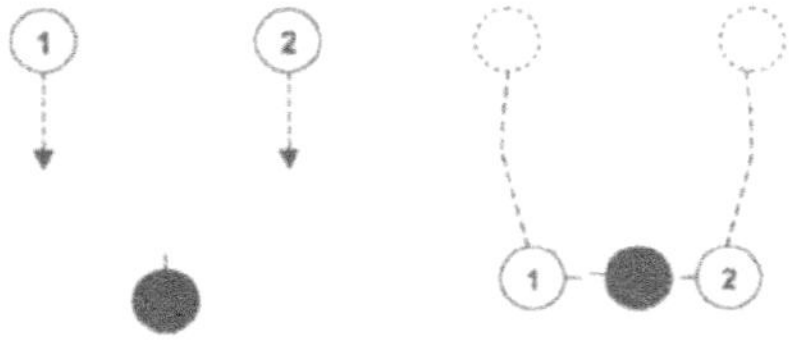

Offensive Strategy: Three Against Two. If three bayonet fighters engage two opponents, the fighters advance together keeping their opponents to the inside.

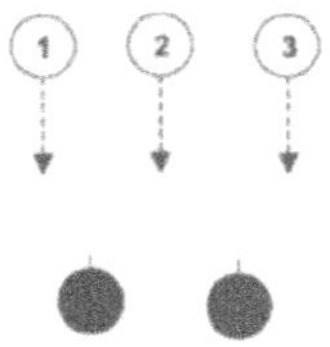

Fighters 1 and 3 engage opponents. Fighter 2 attacks the opponent's exposed flank engaged by fighter 1 and destroys the opponent.

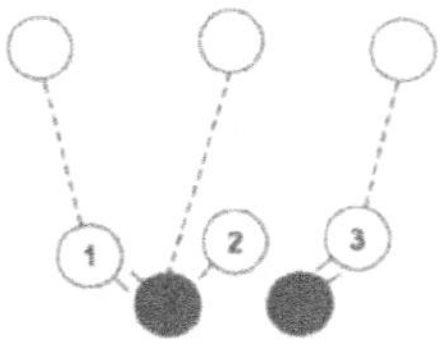

Fighters 1 and 2 turn and attack the exposed flank of the opponent engaged by fighter 3 and destroy the opponent.

Defensive Strategy: One Against Two. If a fighter is attacked by two opponents, the fighter immediately positions himself at the flank of the nearest opponent and keeps that opponent between himself and the other opponent.

Using the first opponent's body as a shield against the second opponent, the fighter destroys the first opponent quickly before the second opponent moves to assist.

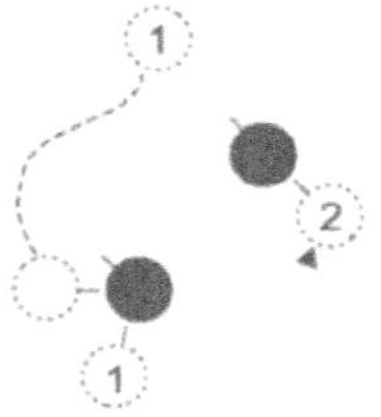

Then, the fighter engages and destroys the second opponent.

Defensive Strategy: Two Against Three. If two fighters are attacked by three opponents, the fighters immediately move to the opponent's flanks.

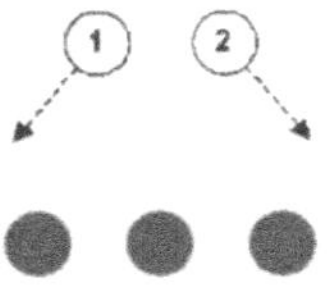

Fighters 1 and 2 quickly attack and destroy their opponents before the third opponent closes in.

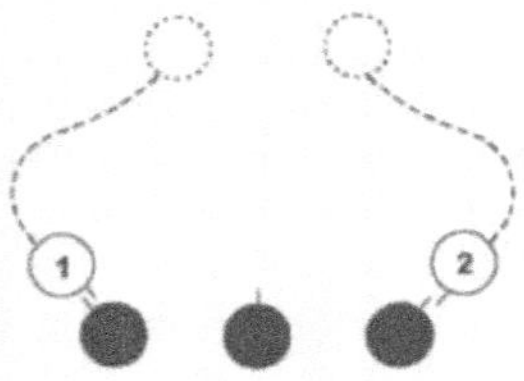

Fighter 1 engages the third opponent while fighter 2 attacks the opponent's exposed flank and destroys the opponent.

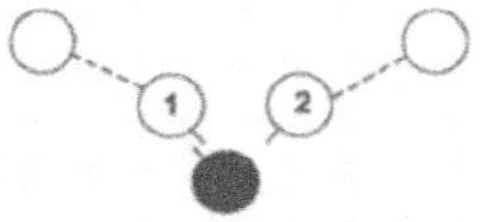

2. Nonlethal Rifle and Shotgun Retention Techniques

Most Marines are armed with the M16A2 service rifle. Marines are taught to keep their weapon with them at all times. Marines must be constantly alert to their surroundings and the people moving in and around their environment. Marines may be confronted by an individual who tries to take theii weapons. If this happens, Marines should not struggle with the individual. To retain positive control of their weapons, Marines must understand and apply weapons retention techniques, otherwise known as armed manipulation. The fol lowing techniques can be used with either the rifle or the shotgun.

Blocking Technique

To execute a blocking technique, Marines-

Stand in a defensive posture.

Use the weapon to block the opponent by thrusting it out firmly, with the elbows bent. Do not try to hit the opponent with the rifle. The rifle is used as a barrier.

Technique if Opponent Grabs Weapon Underhanded

Usually an opponent will try to grab the weapon or block it as an instinctive action. If the opponent uses an underhand grab to seize the handguards of the weapon, Marines—

Trap the opponent's closest finger(s) above the knuckle with the thumb so he cannot release his grip-

Apply bone pressure on the opponent's finger to initiate pain compliance.

Rotate the barrel of the weapon up or down quickly while maintaining pressure on the opponent's hand. At the same time, quickly pivot to off-balance the opponent.

Technique if Opponent Grabs Weapon Overhanded

If the opponent uses an overhand grab to seize the weapon overhanded, Marines—

Trap the opponent's finger to hold his hand in place.

Rotate the barrel to place it across the opponent's forearm and apply downward pressure. This ac-

tion is similar to an armbar.

Technique if the Opponent Grabs the Muzzle

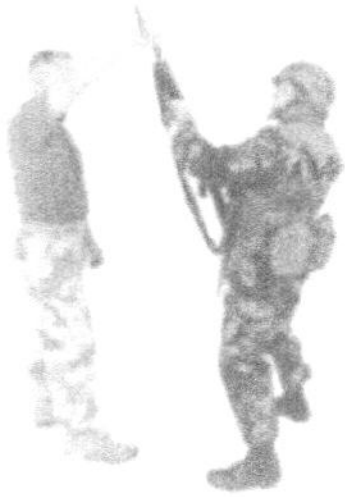

If the opponent grabs the muzzle of the weapon, Marines—

Rotate the muzzle quickly in a circle motion, Slash downward with the muzzle to release the opponent's grip.

Butt Strikes

Strikes with the butt of the weapon control or ward off an attacker. During any of the retention techniques, Marines use the heel or cutting edge of the weapon to deliver butt strikes to the inside or the outside of the opponent's thigh. The inside butt strike targets the femoral nerve as illustrated in the photo to the left.

The outside butt strike targets the peroneal nerve as illustrated in the photo to the left. Strikes can be made to the outside or inside of the thighs. If a strike to one side of the thigh misses, Marines follow back through with the butt of the weapon on the other side of the thigh.

Off-balancing Techniques

Marines apply off-balancing techniques to throw an opponent to the ground and retain possession of the weapon.

If the opponent grabs the weapon and pushes, Marines should not push on the weapon. They should—

Move with the momentum and movement of the

opponent by pivoting in the direction of the movement by stepping back.

Throw the opponent to the ground with a quick jerking movement by lowering the mu/./le and swinging the butt of the weapon.

If the opponent grabs the weapon and pulls, Marines—

Step on the opponent's foot and push forward to off-balance him and drive him to the ground.

Sweep the opponent's feet out from under him by hooking his leg with the leg and kicking backward.

3. Nonlethal Handgun Retention Techniques

Many Marines are armed with the M9 service pistol. Marines must keep their weapons in their possession at all times. Marines must be constantly alert to their surroundings and the people moving in and around the environment. Marines may be confronted by an individual who tries to take their weapons. To retain positive control of the weapon, Marines must understand and apply handgun retention techniques.

Blocking Technique

Marines perform the following blocking techniques if an opponent attempts to grab their pistol in the holster. Marines—

Place the body between the weapon and the opponent by immediately pivoting so the weapon is away from the opponent.

Step back and away from the opponent while placing the hand on the pistol grip.

Extend the left hand and block, deflect, or strike the opponent's arm.

Armbar Technique

Marines use the armbar technique when an opponent uses his right hand to grab their pistol in the holster. To execute the armbar technique, Marines—

Trap the opponent's right hand by grasping the opponent's wrist or hand with the right hand and applying pressure against the body.

Step back with the right foot and pivot sharply to the right to be next to the opponent. Always pivot in a direction that keeps the weapon away from the opponent.

Straighten the opponent's arm to apply an armbar. The arm should be straight across the torso.

Continue pivoting to the right while pulling back on the opponent's shoulder. This action may break the opponent's arm.

Place the left hand across the opponent's face and apply pressure back and down to take the oppo-

nent to the ground. Pressure applied beneath the nose or on the trachea is equally effective.

Wristlock Technique

Marines use the wristlock technique when an opponent grabs their pistol while it is in the holster with his right hand. To execute the w ristlock technique, Marines—

Grasp the opponent's wrist or hand with the right hand and apply pressure against the body.

Step back and away from the opponent and pivot to the right so the weapon is away from the opponent. Always pivot in a direction that keeps the weapon away from the opponent.

Reach over the opponent's arm with the left hand and grab his hand, applying pressure against his arm with the left forearm. Execute a wristlock.

Incorporate the second hand into the wristlock and, stepping back with the left foot, pivot to the left.

Execute a two-handed wristlock by exerting downward pressure with the thumb and rotating his hand to the left.

Continue to pivot to off-balance the opponent and drive him to the ground.

Note: If the opponent grabs the pistol with his left hand, Marines execute the wristlock with one hand and step in toward the opponent, rather than

away from the opponent.

Softening Techniques

Handgun retention techniques use softening techniques applied to pressure points. Bone pressure

and strikes with the hands (i.e., hammer fist), knees, and feet are also effective softening techniques.

Pressure Points. Marines use pressure point techniques to get the opponent to loosen his grip. Marines use their finger tips to apply pressure to the webbing between the index finger and thumb, the jugular notch, and the brachial plexus tie in. The following figure illustrates pressure applied to the brachial plexus tie in.

Bone Pressure. Bone pressure is the application of pressure on a bone against a hard object to initiate pain compliance. To apply bone pressure, Marines use their hand to trap the opponent's hand on the weapon. Marines apply a slow, steady pressure to the opponent's hand and fingers until his grip is softened or he releases his hold.

Strikes. If it is difficult to apply a retention technique, Marines employ strikes or kicks to force the opponent to loosen his grip. Strikes to the

eyes, the arms (radial nerve), or shoulder (brachial plexus tic in) soften the opponent's grip on the weapon.

Kicks and knee strikes to the peroneal nerve, the femoral nerve, or the groin arc effective because the opponent is typically unprepared to counter the strike.

Stomping on the top of the opponent's foot may distract him or loosen his grip on the weapon.

4. Firearm Disarmament Techniques

Marines use firearm disarmament techniques during a close-range confrontation if they are un-

armed and the opponent has a firearm (pistol). These techniques are equally effective if Marines are armed but do not have time to withdraw and present the weapon. The goal of firearm disarmament techniques is to gain control of the situation

so Marines gain the tactical advantage. The goal is not necessarily to get control of the opponent's weapon.

Pistol to the Front

This technique is used when Marines are unarmed and the opponent has a pistol pointing at their front (e.g., head, chest). The technique is the same if the opponent sticks the pistol under the Marine's chin. To execute the counter when an opponent is pointing a pistol toward the front of a Marine, Marines—

Place the hands close to the weapon, about chest high, palms out.

Use the left hand to grab the opponent's forearm and push the opponent's hand with the pistol away to clear the body from in front of the weapon. At the same time, rotate the right shoulder back to clear the body from the weapon.

Maintain control of the opponent's arm.

Grasp the weapon with the right hand by placing the thumb underneath the pistol and the fingers over top of the pistol.

Keep the right hand wrapped tightly around the muzzle and quickly rotate the pistol in the opponent's hand so the muzzle is facing the opponent.

Grasp and pull the opponent's wrist or forearm away from the body while rotating the weapon.

Rotate the weapon toward the opponent while pulling it up and back and out of the opponent's grasp.

Pistol to the Rear

This technique is performed when Marines are unarmed and the opponent has a pistol pointing to the back of the Marine's head. To execute the counter when the opponent is pointing a pistol to the rear, Marines—

Place the hands close together about chest high, palms out.

Step back with the left foot, pivoting on the right foot so the side is against the opponent's front. This action clears the body from the weapon's line of fire. Keep the left hand up.

Pivot on the left foot to face the opponent and, at the same time, raise the left elbow and reach over the top of the opponent's arm with the left arm.

Wrap the left arm tightly around the opponent's arm above his elbow to control it.

Push on the opponent's shoulder with the right hand while pulling up with the left arm to achieve an armbar. This action releases the opponent's grip on the weapon. If necessary, the opponent can be taken to the ground with a leg sweep.

Note: To execute this technique, the weapon must be close to or touching the rear of the Marine. If the weapon is too far away from the body, this technique would be difficult to execute or it would be ineffective.

CHAPTER 3

HAND-HELD WEAPONS

This chapter describes all techniques for a right-handed person. However, all techniques can be executed from either side.

In drawings, the Marine is depicted in woodland camouflage utilities; the opponent is depicted without camouflage. In photographs, the Marine is depicted in woodland camouflage utilities; the opponent is depicted in desert camouflage utilities.

Marines must know how to defend against attacks when an opponent is either unarmed or armed with a held-held weapon. This chapter addresses the combative use of knives, specific weapons of opportunity, and sticks. However, virtually anything can be used as a hand-held weapon.

1. Fundamentals of Knife Fighting

Marines must be trained in knife fighting techniques. Marines experienced in offensive knife techniques can cause enough damage and massive trauma to stop an opponent. When engaged against each other, experienced knife fighters employ various maneuvers and techniques that are specific to knife fighting. Seldom, if ever, will Marines engage an opponent in a classical knife fight.

Note: When armed with a rifle, Marines are issued a bayonet. When armed with a pistol, Marine are issued a combat knife.

Angles of Attack

There are six angles from which an attack with a knife can be launched:

- Vertical strike coming straight down on an opponent.
- Forward diagonal strike coming in at a 45-degree angle to the opponent.
- Reverse diagonal strike coming in at a 45-degree angle to the opponent.
- Forward horizontal strike coming in parallel to the ground.
- Reverse horizontal strike coming in parallel to the ground.
- Forward thrust coming in a straight line to the opponent.

Target Areas of the Body

During any confrontation, the parts of the opponent's body that are exposed or readily accessible will vary. The goal in a knife fight is to attack the body's soft, vital target areas that are readily accessible (e.g, the face, the sides and front of the neck, the lower abdomen [or groin]).

Neck. Carotid arteries, located on either side of the neck, are good target areas because they are not covered by body armor or natural protection.

Lower Abdomen (or Groin). The lower abdomen (or groin region) is a good target area because it is not covered by body armor.

Heart. The heart, if not covered by body armor, is an excellent target which, if struck, can prove fatal in a matter of seconds or minutes.

Secondary Targets. There are secondary target areas that will cause substantial bleeding if an artery is severed. These target areas are not immediately fatal, but can become fatal if left unattended. Attacks to—

The legs can cause a great deal of trauma and prove fatal. For example, the femoral artery located on the inside of the thigh is a large artery which, if cut, will cause extensive blood loss.
The brachial artery, located between the biceps and triceps on the inside of the arm, can cause extensive bleeding and damage.
The arm's radial and ulnar nerves can cause extensive bleeding and damage.

Movement

Marines can move anywhere within a 360-degree circle around the opponent. This allows accessibility to different target areas of the opponent's body. Marines should avoid being directly in front of an opponent because the opponent can rely on his forward momentum to seize the tactical advantage. If Marines face an opponent, movement is made in a 45-degree angle to either side of the opponent. This angle avoids an opponent's strike and places Marines in the best position to attack an opponent.

Wearing the Combat Knife

Marines must wear the combat knife where it is easily accessible and where it can best be retained. It is recommended the combat knife be worn on the weak side hip, blade down, sharp edge facing forward. Marines can place it behind the magazine pouch where it is easily accessible to them, but not easily grabbed by an opponent.

Grip

The grip on the knife should be natural. Marines grasp the knife's grip with the fingers wrapped around the grip naturally as it is pulled out of its sheath. This is commonly known as a hammer grip. The blade end of the knife is always facing the opponent.

Stance

Marines use the basic warrior stance as the foundation for knife techniques. The left hand is a vertical shield that protects either the ribs or the head and neck. The right elbow is bent with the blade pointing forward toward the opponent' s head. This position serves as an index point, where all techniques are initiated.

Principles of Knife Fighting

The following are key principles of knife fighting:

Execute movements with the knife blade within a box, shoulder-width across from the neck down to the waistline. The opponent has a greater chance of blocking an attack if the blade is brought in a wide, sweeping movement to the opponent.
Close with the opponent, coming straight to the target.
Move with the knife in straight lines.
Point the knife's blade tip forward and toward the opponent.
Apply full body weight and power in each of the knife techniques. Full body weight should be put into the attack in the direction of the blade's movement (slash or thrust).
Apply constant forward pressure with the body and blade to keep the opponent off-balanced.

2. Knife Fighting Techniques

Slashing Techniques

Marines use slashing techniques to close with an enemy. Slashing techniques distract the opponent or damage the opponent so Marines can close in. Typically, Marines target the opponent's limbs, but any portion of the body that is presented can become a target.

Vertical Slash Technique. The vertical slash follows a vertical line straight down through the target. To execute the vertical slash, Marines—

Thrust the right hand out and bring the weapon straight down on the opponent, continuing to drag the knife down through the opponent's body.

Maintain contact on the opponent's body with the blade of the knife.

Forward Slash Technique. The forward slash follows a straight line in a forehand stroke, across the target areas of either the neck (high diagonal slash) or abdominal region (low horizontal slash). To execute the forward slash, Marines-
Extend the right hand while simultaneously rotating the palm up until the knife blade makes contact with the opponent.

Snap or rotate the wrist through the slashing motion to maximize blade's contact with the opponent.

Drag the knife across the opponent's body, from right to left, in a forehand stroke. The movement ends with the forearm against the body and the knife at the left hip with its blade oriented toward the opponent.

Reverse Slash Technique. The reverse slash is a follow-up technique to a forward attack. It allows Marines both a secondary attack and the ability to resume the basic warrior stance. The reverse slash follows a straight line in a backhand stroke, across the target areas of either the neck (high diagonal slash) or abdominal region (low horizontal slash). To execute the reverse slash, Marines—

Extend the right hand while simultaneously rotating the palm down until the knife blade makes contact with the opponent.

Snap or rotate the wrist through the slashing motion to maximize the blade's contact with the opponent.

Drag the knife across the opponent's body, from left to right, in a backhand stroke. Maintain contact on the opponent's body with the blade of the knife.

Thrusting Techniques

The primary objective of knife fighting is to insert the blade into an opponent to cause extensive damage and trauma. This is done with a thrusting technique. Thrusting techniques are more effective than slashing techniques because of the damage they can inflict. However, Marines use slashing techniques to close with the enemy so that they are closer to the opponent, which allows them to use the thrusting technique.

Vertical Thrust. The thrusting motion follows a vertical line straight up through the target (low into the abdomen region or high into the neck). To execute the vertical thrust, Marines—

Thrust the right hand toward the target, inserting the knife blade straight into the opponent.

Pull the knife out of the opponent.

Forward Thrust. The forward thrust follows a straight line straight into the opponent's neck (high thrust) or abdominal region (low thrust). To execute the forward thrust, Marines—

Thrust the right hand, palm down, toward the target, inserting the knife blade straight into the opponent.

Rotate the palm up once the knife is inserted to twist the blade.

Drop the right elbow and bring the knife to the opposite side of the opponent's body from where it was inserted. At the same time, rotate the hips and shoulders downward to bring body weight to bear on the attack.

Reverse Thrust. The reverse thrust is a follow-up technique to a forward attack. It allows Marines both a secondary attack and the ability to resume the basic warrior stance. The reverse thrust follows a horizontal line straight into the opponent's neck (high thrust) or abdominal region (low thrust). To execute the reverse thrust, Marines-

Bend the right arm, crossing the arm to the left side of the body.

Thrust the right hand, palm up, toward the target, inserting the knife blade straight into the opponent.

Rotate the palm down to twist the blade once the knife is inserted.

Bring the knife to the opposite side of the opponent's body from where it was inserted.

3. Weapons of Opportunity

During an unarmed close combat situation, Marines use their bodies as weapons, but they should be ready and able to use anything around them as a weapon. For example, Marines could throw sand or liquid in an opponent's eyes to temporarily impair his vision or smash the opponent's head with a rock or helmet. Marines must use whatever means are available and do whatever it takes to take control of the situation and to win, or they face the possibility of losing their lives. Some weapons of opportunity are discussed in the following subparagraphs.

Entrenching Tool

An entrenching tool (E-tool) is commonly carried by Marines. It can be an excellent weapon, especially when sharpened. Marines can use the E-tool to block, slash, and thrust at an opponent.

Tent Pole and Pins

Marines can use tent poles and pins to block, strike, or thrust at an opponent.

Web Belt

Marines can stretch a web belt between their hands to block attacks by an opponent.

Battlefield Debris

Marines can use debris on the battlefield (e.g., sticks, glass, a sharp piece of metal) to cut, slash, or stab an opponent. They can also use other types of debris such as shovel or ax handles, boards, metal pipes, or broken rifles to strike an opponent or apply a choke.

Helmet

A helmet can be used to strike an opponent on an unprotected area like the head and face. Grasp the rim of the helmet and thrust the arms forward, striking the opponent with the top of the helmet.

4. Fundamentals of Combative Stick

On the battlefield, Marines must be ready and able to use anything as a weapon. They must learn and be able to use techniques that can be employed with most weapons of opportunity. Among these techniques are combative stick techniques. Combative stick techniques can be used with a stick, a club, a broken rifle, an E-tool, or even a web belt.

Angles of Attack

There are six angles from which an attack with a hand-held weapon can be launched:

Vertical strike coming straight down on an opponent.
Forward diagonal strike coming in at a 45-degree angle to the opponent.
Reverse diagonal strike coming in at a 45-degree angle to the opponent.
Forward horizontal strike coming in parallel to the ground.
Reverse horizontal strike coming in parallel to the ground.
Forward thrust coming in a straight line to the opponent.

Grip

Grasp the stick about 2 inches from its base.

Stance

The basic warrior stance serves as the foundation for combative stick techniques. The left hand becomes a vertical shield that protects the ribs or the head and neck. Depending on how heavy the weapon is, it should be held at a level approximately shoulder height.

Movement

Movement during combative stick techniques is the same as it is for other close combat techniques. Marines can move anywhere within a 360-degree circle around the opponent. This allows accessibility to different target areas of the opponent's body and gains a tactical advantage. Marines should avoid being directly in front of an opponent because the opponent can rely on his forward momentum to seize the tactical advantage. If Marines face an opponent, movement is made in a 45-degree angle to either side of the opponent. This angle avoids an opponent's strike and places Marines in the best position to attack an opponent.

5. Combative Stick Techniques

Strikes

Strikes are intended to inflict as much damage on an opponent as possible. Striking techniques apply to a weapon of opportunity such as a stick, a tent pole, a club, a broken rifle, an E-tool, or a Pipe-

Vertical Strike. To execute the vertical strike, Marines—

Bend the right arm, extending the weapon over the back of the right shoulder.

Rotate the forearm straight down off the elbow to bring the weapon down on the opponent.

Rotate the hips and shoulders forcefully toward the opponent.

Follow through with the strike by allowing the weight of the weapon to go through the target area of the body.

Forward Strike. To execute a forward strike, Marines—

Step forward with the left foot in the direction of the strike.

Bend the right arm with the elbow extending out to the right and the weapon extended over the right shoulder.

Rotate the forearm to the right of the elbow to bring the weapon down onto the opponent. At the same time, forcefully rotate the hips and shoulders toward the opponent.

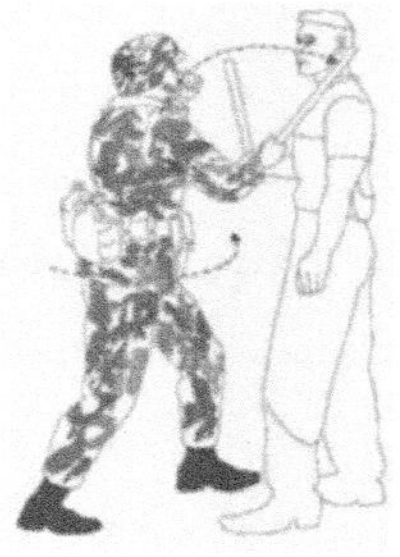

Follow through with the strike by allowing the weight of the weapon to go through the target area of the body.

Reverse Strike. The reverse strike is a follow-up technique to a forward strike. It allows Marines both a secondary attack and the ability to resume the basic warrior stance.To execute a reverse strike, Marines—

Step forward with the right foot in the direction of the strike.

Bend the right arm with the hand near the left shoulder. The weapon is extended over the left shoulder.

Rotate the forearm to the right of the elbow to bring the weapon down onto the opponent. At the same time, forcefully rotate the hips and shoulders toward the opponent.

Follow through with the strike by allowing the weight of the weapon to go through the target area of the body.

Forward Thrust. To execute the forward thrust, Marines—

Grasp the stick with the left hand, palm up, in a position where the stick can be controlled with two hands.

Lift the left leg and lunge forward off the ball of the right foot. At the same time, thrust the end of the weapon directly toward the opponent by thrusting both hands forward in a straight line.

6. Blocking Techniques

A block is meant to deter or deflect an attack by an opponent. A block sets up Marines for a follow-on attack against the opponent. Blocks are executed by deflecting, rather than hitting or following through like a strike.

Blocks Against Unarmed Attacks

To block against an unarmed attack, Marines—

Step forward at a 45-degree angle with either the right or left foot. This moves the body out of the line of attack.

Raise the left arm and block the strike with the meaty portion of the forearm.

Employ the stick by one of two techniques:

Strike the opponent with the stick.
Use the stick to block on two points of contact. When the stick is used to block, it serves as an incidental strike. This technique is only used if Marines have closed with the opponent and are inside his strike.

Blocks Against Armed Attacks

Block for a Vertical Strike. To execute the block for a vertical strike, Marines—

Step forward with the left foot at a 45-degree angle to the left. This moves the body out of the line of attack.

Block on two points of contact to disperse the impact of the attack:

Block the opponent's stick by positioning the stick so it is perpendicular to the opponent's stick. If the stick is not perpendicular to the opponent's stick, the stick can slide through and make contact on the Marine.
Block the opponent's wrist or forearm with the back of the left forearm.

Note: Use the stick to block the opponent's arm if closer to the opponent. It is the same movement as blocking with the arm, except the opponent's arm is blocked with both the stick and the arm.

Block for a Forward Strike. To execute the block for a forward strike, Marines-

Step forward with the left foot at a 45-degree angle to the left. This moves the body out of the line of attack and inside the opponent's strike.

Block on two points of contact to disperse the impact of the attack:

- Block the opponent's wrist or forearm with the meaty portion of the back of the left forearm.
- Strike the opponent's attacking biceps with the stick.

Block for a Reverse Strike. To execute the block for a reverse strike, Marines—

Step forward at a 45-degree angle with the right foot to the right. This moves the body out of the line of attack and inside the opponent's strike.

Block on two points of contact to disperse the impact of the attack:

Block the opponent's stick by positioning the stick so it is perpendicular to the opponent's stick.

Block the opponent's forearm with the meaty portion of the left forearm.

Note: If closer to the opponent, block his triceps with the back of the left forearm and strike his forearm with the stick.

7. Unarmed Against Hand-Held Weapons

If Marines are engaged against an opponent with a knife, a stick, or some other weapon of opportunity, they must establish and maintain an offensive mindset, not a defensive mindset. Their survival depends on it. They cannot afford to think about getting cut or hurt.

Angles of Attack

Before Marines can learn to block or counter an attack with a hand-held weapon (e.g., knife, stick), they must know from what angle the opponent is attacking. There arc six angles from which an opponent will typically attack with a hand-held weapon:

Vertical strike coming in straight down on the Marine.
Forward diagonal strike coming in at a 45-degree angle to the Marine.
Reverse diagonal strike coming in at a 45-degree angle to the Marine.
Forward horizontal strike coming in parallel to the ground.
Reverse horizontal strike coming in parallel to the ground.
Forward thrust coming in a straight line to the Marine.

Blocks

If the opponent has a hand-held weapon (e.g., knife or stick), Marines parry the opponent's hand or arm to block the attack.

Basic Block. To execute the basic block technique, Marines-

Step forward at a 45-degree angle to move out of the line of the attack. Always step in the direction of the strike.

Thrust the forearms forward, hands up, against the opponent's attacking arm. Contact is made on the opponent's arm with the backs of the forearms.

Block for a Vertical Strike. To execute the block against a vertical strike, Marines—

Step forward with the left foot at a 45-degree angle to the left to move out of the line of attack.

Thrust the forearms forward, hands up, against the outside of the opponent's attacking arm.

Block for a Forward Strike. To execute the block against a forward diagonal or forward horizontal strike, Marines-

Step forward with the left fool inside the opponent's attacking arm.

Block the attack with both arms bent so the forearms make contact with the opponent's biceps and forearm.

Block for a Reverse Strike. To execute the block against a reverse diagonal or reverse horizontal strike, Marines—

Step forward with the right foot to the outside of the opponent's attacking arm.

Block the attack with both arms bent so the forearms make contact with the opponent's triceps and forearm.

Block for a Forward Thrust. To execute the block against a forward thrust, Marines—

Bend at the waist, move the hips backward, and jump backward with both feet to move away from the attack. This action is known as "hollowing out."

Hollow out and block the attack with the arms bent and hands together on top of the attacking arm.

Overlap the hands slightly so one thumb is on top of the other hand's index finger. The other thumb should be under the other hand's index finger.

8. Counters to Hand-Held Weapon Attacks

Principles of Counters

A counter is used to control the situation to regain the tactical advantage and end the fight. Regardless of the type of weapon or angle of attack, the following principles apply to countering the attack with a hand-held weapon:

Move out of the line of attack. Movement is executed in a 45-degree angle forward to the left or right.

- Block the attack.

Note: The first two actions are taken simultaneously.

- Take control of the weapon by controlling the hand or arm that is holding the weapon. Never attempt to grab the opponent's weapon.
- Execute the appropriate follow-up to end the fight; e.g., strikes, joint manipulations, throws, or takedowns (see chap. 8). Marines should continue to attack the opponent until the fight ends.

Counter Techniques

There are two techniques that can be used to counter any armed attack: forward armbar counter and reverse armbar counter. These techniques can be used to counter a vertical attack, a forward diagonal strike, or a forward horizontal strike. With minor variations, the same techniques arc used to counter reverse strikes. A third technique, the bent armbar counter, is used to counter a vertical attack.

Forward Armbar Counter. To execute the forward armbar counter to an attack coming from a forward strike, Marines-

Step forward with the left foot inside the opponent' s attacking arm.

Block the attack with both arms bent so the forearms make contact with the opponent's biceps and forearm.

Slide the left arm over the opponent's forearm and wrap the arm tightly around his arm, trapping the opponent's attacking arm between the biceps and torso.

Place the right hand on the opponent's shoulder or upper arm to further control his arm and to effect an armbar.

Execute an armbar and continue to exert steady pressure against the arm.

Reverse Armbar Counter. To execute the reverse armbar counter to an attack coming from a forward strike, Marines-

Step forward with the left foot inside the opponent' s attacking arm.

Block the attack with both arms bent so the forearms make contact with the opponent's biceps and forearm.

Control the opponent's arm with the left hand and pivot to the right so the back is against the opponent's side. Immediately slide the right arm over the opponent's biceps and wrap the arm tightly around his arm, trapping the opponent's attacking arm between the biceps and torso.

Grasp the opponent's wrist with the left hand and twist his thumb away from the body.

Take the opponent to the ground with an armbar takedown if he does not drop the weapon from the wristlock.

Bent Armbar Counter. This counter is particularly effective against a vertical attack. To execute the bent armbar counter, Marines—

Step forward with the left foot inside the opponent's attacking arm.

Block the attack with both arms bent so the forearms make contact with the opponent's biceps and forearm.

Grasp the opponent's forearm with the left hand. At the same time, slide the right arm underneath the opponent's triceps and grasp the opponent's forearm or wrist with the right hand.

Apply downward pressure with the hands against the opponent's forearm to off-balance the opponent.

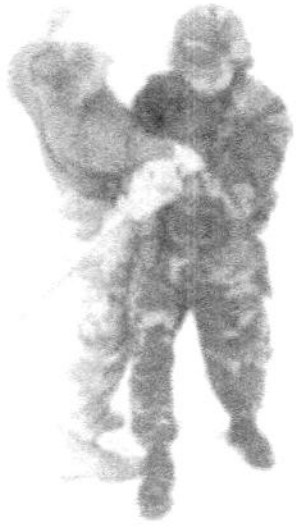

(reverse blank)

CHAPTER 4

STRIKES

This chapter describes all techniques for a right-lmnded person. However, all techniques can be executed from either side.

In drawings, the Marine is depicted in woodland camouflage utilities; the opponent is depicted without camouflage. In photographs, the Marine is depicted in woodland camouflage utilities; the opponent is depicted in desert camouflage utilities.

Strikes are unarmed individual hitting techniques. Strikes use the hands, elbows, knees, feet, and, in some instances, other parts of the body as personal weapons. Marines must know how to execute strikes effectively. They must also know how to counter strikes from an opponent.

1. Principles of Punches

Muscle Relaxation

Muscle relaxation is crucial when executing punches. The natural tendency in a fight is to tense up, which results in rapid fatigue and decreased power generation. Marines who remain relaxed during a close combat situation generate greater speed, which results in greater generation of power. Relaxing the forearms generates speed and improves reaction time. At the point of impact, Marines clench the fist to cause damage to the opponent and avoid injury to the wrist and hand.

Weight Transfer

Weight transfer is necessary to generate power in a punch. Marines accomplish this by-

Rotating their hips and shoulders into the attack.

Moving their body mass straight forward or backward in a straight line.

Dropping their body weight into an opponent. Body mass can be transferred into an attack from high to low or from low to high.

Rapid Retraction

When Marines deliver a punch, rapid retraction of the fist is important. Once the hand has made contact with the target, Marines quickly return to the basic warrior stance. Rapid retraction-

Returns the hand and arm to the protection afforded by the basic warrior stance.

Prevents the opponent from being able to grab the hand or arm.

Permits the hand and arm to be "chambered" or "re-cocked" in preparation for delivering a subsequent punch.

Telegraphing

Telegraphing a strike occurs when body movements inform the opponent of the intention to launch an attack. Staying relaxed helps to reduce telegraphing.

Often, an untrained fighter telegraphs his intention to attack by drawing his hand back in view of his opponent, changing facial expression, tensing neck muscles, or twitching. These movements, however small, immediately indicate an attack is about to be delivered. If the opponent is a trained fighter, he may be able to evade or counter the attack. If the opponent is an untrained fighter, he may still be able to minimize the effect of an attack.

2. Punches

Punches may be thrown during any hand-to-hand confrontation. Most people resort to punching because it is a natural reaction to a threat. The purpose of a punch is to stun the opponent or to set him up for a follow-up finishing technique. However, punches should only be executed to the soft tissue areas of an opponent. A correctly delivered punch maximizes the damage to an opponent while minimizing the risk of injury to Marines.

Basic Fist

Punches are executed using the basic fist. To make the basic fist, the fingers are curled naturally into the palm of the hand and the thumb is placed across the index and middle fingers. Do not clench the fist until movement has begun. This reduces muscular tension in the forearm and increases speed and reaction time. Just before impact, Marines exert muscular tension on the hand and forearm to maximize damage to the opponent and reduce their chances of injury. Contact should be made with the knuckles of the index and middle fingers.

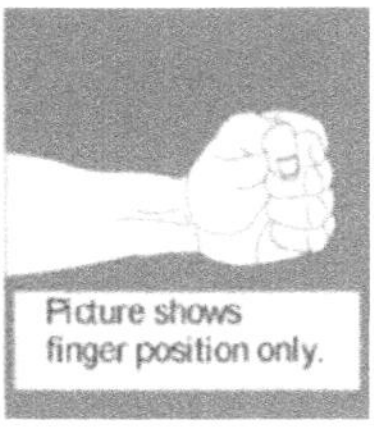

When striking with the basic fist, Marines must keep the hand straight, or in line, with the wrist to avoid injury to the wrist.

Lead Hand Punch

The lead hand punch is a snapping straight punch executed by the forward or lead hand. It is a fast punch designed to keep the opponent away and to set up an attack. A lead hand punch conceals movement and allows Marines to get close to the opponent. Lead hand punches should strike soft tissue areas, if possible. To execute the lead hand punch, Marines-

Snap the lead hand out to nearly full extension, while rotating the palm to the ground.

Contact the opponent with the first two knuckles of the fist.

Retract the hand immediately, resuming the basic warrior stance.

Rear Hand Punch

The rear hand punch is a snapping punch executed by the rear (right) hand. It is a power punch designed to inflict maximum damage on the opponent. Its power comes from pushing off the rear leg and rotating the hips and shoulders. To execute the rear hand punch, Marines—

Rotate the hips and shoulders forcefully toward the opponent and thrust the rear hand straight out, rotating the palm down, to nearly full extension.

Shift body weight to the lead foot while pushing off on the ball of the rear foot.

Contact the opponent with the first two knuckles of the fist.

Retract the hand immediately.

Uppercut

The uppercut is a powerful punch originating below the opponent's line of vision. It is executed in an upward motion traveling up the centerline of the opponent's body. It is delivered in close and usually follows a preparatory strike that leaves the target area unprotected. When delivered to the chin or jaw, the uppercut can render an opponent unconscious, cause extensive damage to the neck, or sever the tongue. To execute the uppercut, Marines—

Bend the arms, rotating the palm inboard. The distance the arms bend depends on how close the opponent is.

Rotate the hips and shoulders forcefully toward the opponent, thrusting the fist straight up toward the opponent's chin or jaw.

Contact the opponent with the first two knuckles of the fist.

Retract the hand immediately.

Hook

The hook is a powerful punch that is executed close in and is usually preceded by a preparatory strike. To execute the hook, Marines—

Thrust the right arm in a hooking motion toward the opponent, keeping the elbow bent while forcefully rotating the right shoulder and hip toward the opponent.

Contact the opponent with the first two knuckles of the fist. Continue rotating the shoulder and hip, following through with the fist to the target.

Retract the hand immediately.

3. Strikes with the Upper Body

Strikes stun the opponent or set him up for a follow-up finishing technique. The hands, forearms, and elbows are individual weapons of the arms that can be used to execute strikes including the hammer fist, chin jab, knife hand, eye gouge, and elbow strikes. These strikes provide a variety of techniques that can be used in any type of close combat situation.

Principles of Execution

There are several principles of execution that ensure a strike's effectiveness.

Generating Power. Weight transfer is necessary to generate maximum power in a punch. Marines accomplish this by-

- Rotating the hips and shoulders into the attack.
- Moving the body mass straight forward or backward in a straight line.
- Dropping body weight into an opponent. Body mass can be transferred into an attack from high to low or from low to high.

Muscular Tension. The arms are relaxed until the moment of impact. At the point of impact, Marines apply muscular tension in the hand and forearm to maximize damage to the opponent and to avoid injury to the hand. The arms are relaxed until the moment of impact.

Hit and Stick and Follow-Through. A strike should be delivered so that the weapon (e.g., hand, elbow) hits and remains on the impact site (target) and follows through the target. This technique inflicts maximum damage on the opponent.

Strikes with the arms are executed with "heavy hands"; i.e., the strike is executed by driving through with the strike to allow the weight of the hand to go through the target area of the body. Contact with the opponent should be made with the arm slightly bent, and the arm extends as it moves through the target. This technique allows Marines to deliver strikes effectively without executing full force.

Movement. Movement puts Marines in the proper position for launching an attack against an opponent as well as providing protection.

Movement is initiated from the basic warrior stance and ends by resuming the basic warrior stance. Strikes can be performed with either the left or right arm depending upon—

The angle of attack.
The position of the opponent.
The opponent's available, vulnerable target areas.

Target Areas of the Body

For each strike, there are target areas of the body that, when struck, maximize damage to an opponent. Strikes use gross motor skills as opposed to fine motor skills. The target areas of the body are just that—areas. Pinpoint accuracy on a specific nerve is not needed for the strike to be effective.

Hammer Fist

Striking with the hammer fist concentrates power in a small part of the hand, which, when transferred to the target, can have a devastating effect. The striking surface of the hammer fist is the meaty portion of the hand below the little finger. To execute the hammer fist strike, Marines-

Make a fist and bend the arm at approximately a 45- to 90-degree angle. At the same time, rotate the right hip and right shoulder backward.

Thrust the fist forward onto the opponent while rotating the right hip and shoulder forward.

Rotate the wrist so the hammer fist makes contact on the opponent.

Chin Jab

The chin jab can immediately render an opponent unconscious and cause extensive damage to the neck and spine. The striking surface is the heel of the palm of the hand. To execute the chin jab, Marines—

Bend the right wrist back at a 90-degree angle with the palm facing the opponent and the fingers pointing up.

Keep the right arm bent and close to the body. Extend the hand into a concave position with the fingers slightly spread apart.

Step forward with the left foot toward the opponent, keeping the feet approximately shoulder-width apart and the knees bent. This is done to close with the opponent.

Keep the right arm bent and close to the side. Thrust the palm of the hand directly up under the opponent's chin. At the same time, rotate the right hip forward to drive the body weight into the attack to increase the power of the strike. The attack should travel up the centerline of the opponent's chest to his chin.

Knife Hand

The knife hand is one of the most versatile and devastating strikes. The striking surface is the cutting edge of the hand, which is the meaty portion of the hand below the little finger extending to the top of the wrist. The striking surface is narrow, allowing strikes on the neck between the opponent's body armor and helmet. The knife hand strike is executed from one of three angles: outside, inside, and vertical.

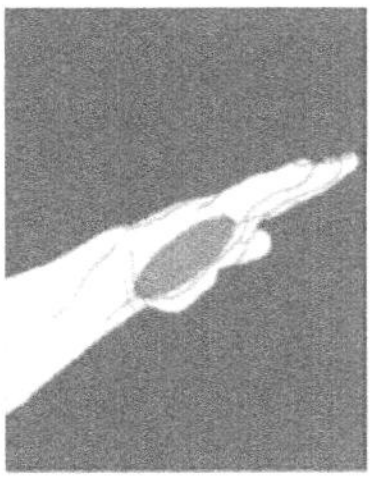

Outside Knife Hand. To execute the outside knife hand strike, Marines—

Execute a knife hand by extending and joining the fingers of the right hand and placing the thumb next to the forefinger (like saluting).

Retract the right hand. At the same time, rotate the right hip and right shoulder backward.

Thrust the knife hand forward (horizontally) onto the opponent while rotating the right hip and shoulder forward.

Inside Knife Hand. To execute the inside knife hand strike, Marines—

Execute a knife hand.

Bring the right hand over the left shoulder. At the same time, rotate the right shoulder forward and the left hip forward.

Thrust the knife hand forward (horizontally) onto the opponent while rotating the right hip and shoulder forward and the left shoulder backward.

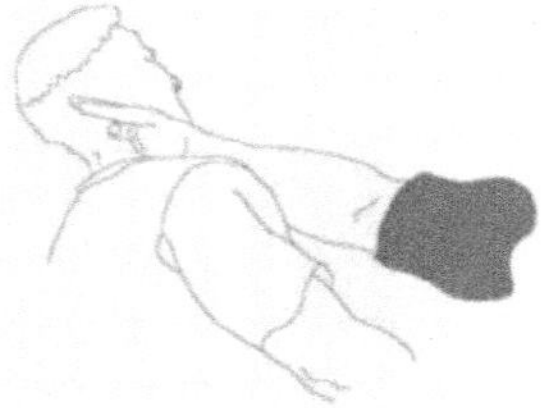

Vertical Knife Hand. When thrown vertically, the knife hand strike comes straight down in a straight line.

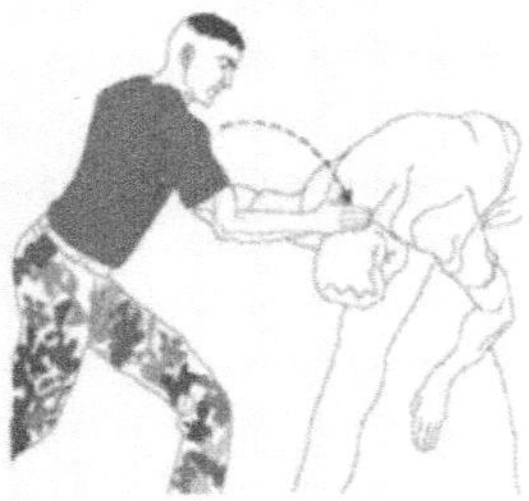

Eye Gouge

The eye gouge is used to attack an opponent's eyes, blinding him so follow-up strikes can be executed. The striking surface is the tips of the fingers or thumb. To execute the eye gouge, Marines—

Extend the right hand with the fingers slightly spread apart to allow entry into the eye sockets.

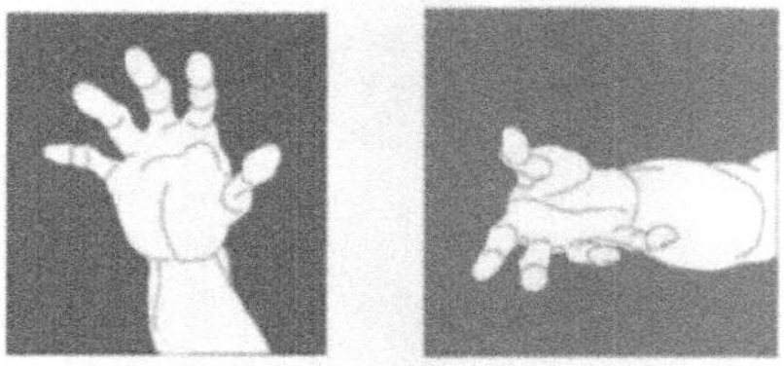

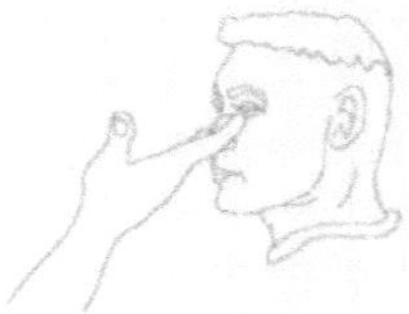

Place the palm of the hand either toward the ground or toward the sky and thrust the right hand forward into the opponent's eyes.

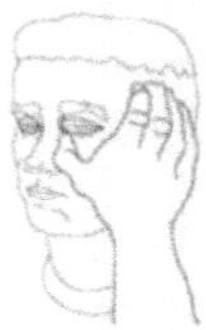

Thrust the hand forward at the opponent's nose level so the fingers or thumb slide naturally into the grooves of the opponent's eye sockets.

Elbow Strikes

The elbow is a powerful weapon that can be used in several different ways to attack virtually any part of an opponent's body. Elbow strikes can be performed either vertically (upward or downward) or horizontally (forward or reverse). The striking surface is 2 inches above or below the point of the elbow, depending upon the angle of attack, the opponent's attack angle, and the position of the opponent.

Above Elbow

Below Elbow

Vertical Elbow Strike (Up). To execute an upward vertical elbow strike, Marines-

Bend the right elbow, keeping the fist close to the body. The fist is at shoulder level and the elbow is next to the torso.

Thrust the elbow vertically upward toward the opponent while rotating the right shoulder and hip forward to generate additional power.

Contact the opponent with the right forearm 2 inches above the point of the elbow.

Vertical Elbow Strike (Down). To execute a downward vertical elbow strike, Marines—

Bend the right elbow, keeping the fist close to the body. The fist is on the shoulder and the elbow is raised well above the shoulder.

Thrust the elbow vertically downward toward the opponent while dropping body weight into the attack to generate additional power.

Contact the opponent with the right triceps 2 inches above the point of the elbow.

Horizontal Elbow Strike (Forward). To execute the forward horizontal elbow strike, Marines-

Tuck the right fist near the chest with the palm heel facing the ground.

Thrust the right elbow horizontally forward toward the opponent. The forearm is parallel to the ground.

Rotate the right shoulder and hip forward to generate additional power.

Contact the opponent with the right forearm 2 inches below the point of the elbow.

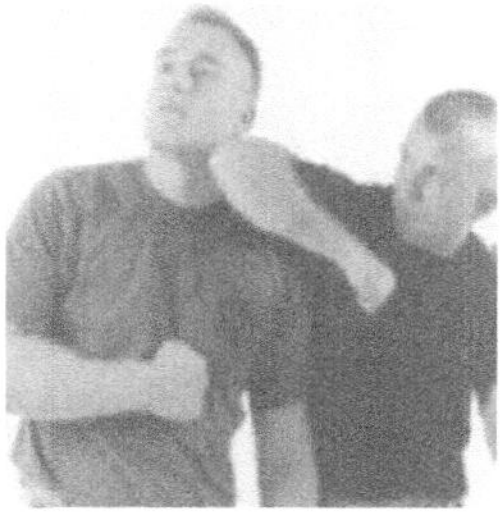

Horizontal Elbow Strike (Rear). To execute the rear horizontal elbow strike, Marines-

Tuck the right fist near the left shoulder with the palm heel facing the ground. At the same time, rotate the right shoulder forward and the left hip forward.

Thrust the right elbow horizontally rearward toward the opponent. The forearm is parallel to the ground and the hand moves toward the direction of the attack.

Rotate the right hip back and the right shoulder backward to generate additional power.

Contact the opponent with the right triceps 2 inches above the point of the elbow.

4. Strikes with the Lower Body

The legs are the body's most powerful weapons because they use the largest muscle groups to generate a strike. Legs are also less prone to injury. The feet are the preferred choice for striking because they are protected by boots. Marines use their feet, heels, and knees to execute kicks, knee strikes, and stomps.

Kicks

The purpose of a kick is to stop an opponent's attack or to create an opening in his defense in order to launch an attack. Kicks can be performed with the left (lead) leg or the right (rear) leg. Kicks with the rear leg have greater power because the hips are rotated into the attack. However, the rear leg is further away from the opponent so a strike with the rear leg will not contact the opponent as quickly as a strike with the lead leg.

Front Kick. The front kick is executed when the opponent is in front of the Marine. The front kick, delivered with either the rear or lead leg, is effective for striking below the waist. Attempting to kick higher results in diminished balance and provides the enemy with a greater opportunity to grab the leg or foot. The striking surfaces arc the toe of the boot or the bootlaces. To execute the front kick, Marines-

Raise the left knee waist high, pivot the hips into the attack, and thrust the left foot forward toward the opponent.

Contact the opponent with the toe of the left boot or bootlaces.

Return to the basic warrior stance.

Side Kick. The side kick, delivered with the lead leg, is effective for striking the knees. The side kick is executed when the opponent is to the side of the Marine. The striking surface is the outside cutting edge of the boot near the heel. To execute the side kick, Marines—

Raise the right knee waist high and rotate the right hip forward.

Thrust the right foot to the right side toward the opponent, turning the foot at a 90-degree angle to maximize the striking surface on the opponent.

Contact the opponent with the cutting edge of the right boot.

Return to the basic warrior stance.

Knee Strikes

Knee strikes are excellent weapons during the close range of close combat fighting. The knee strike is generally delivered in close.

Vertical Knee Strike. The striking surface is the thigh, slightly above the knee. To execute the vertical knee strike, Marines-

Raise the right knee and drive it up forcefully into the opponent. Power is generated by thrusting the leg upward.

Contact the opponent 2 inches above the right knee.

Horizontal Knee Strike. The horizontal knee strike is executed with the leg generally parallel with the ground while rotating the hips to generate power. It is often delivered to the peroneal nerve. The striking surface is the front of the leg, slightly above or below the knee. To execute the horizontal knee strike, Marines-

Raise the right knee, rotate the right hip forward while pivoting on the left foot, and drive the knee horizontally into the opponent.

Contact the opponent 2 inches above the right knee.

Stomps

Stomps are delivered with the feet, usually when the opponent is down. Remember, when the opponent is down, Marines take whatever target is available.

Vertical Stomp. The vertical stomp allows Marines to remain upright and balanced, to rapidly deliver multiple blows with either foot, and to quickly and accurately attack the target. It is the preferred stomp. The striking surface is the flat bottom of the boot or the cutting edge of the heel. To execute the vertical stomp, Marines—

Raise the right knee above the waist with the right leg bent at approximately a 90-degree angle.

Drive the flat bottom of the right boot or the cutting edge of the right heel down onto the opponent forcefully. At the same time, bend the left knee slightly to drop the body weight into the strike.

Ax Stomp. The striking surface of the ax stomp is the cutting edge of the heel. To execute the ax stomp, Marines-

Raise the right heel above the waist, keeping the right leg straight.

Drive the cutting edge of the right heel down onto the opponent forcefully. At the same time, bend the left knee slightly to drop the body weight into the strike.

5. Counters to Strikes

In a close combat situation, an opponent will attempt to strike Marines with punches and kicks. When an opponent uses a strike, Marines must first avoid the strike. This is accomplished by moving quickly and blocking. Next, Marines must get into an offensive position. This allows Marines to use offensive strikes to attack the opponent. Regardless of the strike, the counter to a strike requires Marines to move, block, and strike.

Move

The first step in countering a strike is to move out of the way of the strike's impact. Movement removes Marines from the opponent's intended strike point and positions Marines to attack. Movement is executed at approximately a 45-degree angle to the front or rear. Movement is always initiated from the basic warrior stance. Return to the basic warrior stance with the toe of the lead foot pointing toward the opponent once the movement is complete.

Block

Different blocks are executed based on the strike. These will be covered by individual counters.

Strike

Any of the upper body or lower body strikes can be executed as a follow-on attack or part of the counter to an opponent's strike. The follow-on strike is determined by the angle to the opponent, the position of the opponent, and the opponent's available vulnerable target areas.

Counters to Punches

Counter to a Lead Hand Punch. This counter is used when the opponent throws a lead hand punch. To execute the counter to the lead hand punch, Marines-

Step forward and to the left at approximately a 45-degree angle, moving in to the outside of the opponent's attacking arm.

Raise the left arm and block or deflect the opponent's lead hand with the palm of the hand or the meaty portion of the forearm.

Hit and stick by leaving the left arm against the opponent's right arm while stepping forward and to the left at approximately a 45-degree angle to close with the opponent.

Execute a finishing technique, such as a strike or a kick, to the opponent's exposed target areas.

Counter to a Rear Hand Punch. This counter is used when the opponent throws a rear hand punch. To execute the counter to the rear hand punch, Marines—

Step forward and to the left at approximately a 45-degree angle, moving in to the outside of the opponent's attacking arm.

Raise the left arm and block or deflect the opponent' s rear hand with the palm of the hand or the meaty portion of the forearm.

Hit and stick by leaving the left arm against the opponent's right arm while stepping forward and to the right at approximately a 45-degree angle to close with the opponent.

Execute a finishing technique, such as a strike or a kick, to the opponent's exposed target areas.

Counters to Kicks

Counter to a Front Kick (Left or Lead Leg). This counter is used when the opponent executes a front kick with his left leg. To execute the counter to a front kick, Marines-

Step forward and to the right at approximately a 45-degree angle, moving in to the outside of the opponent's striking leg.

Lower the left arm and block or deflect the opponent' s leg with the palm of the hand or the meaty portion of the forearm.

Hit and stick by leaving the left arm against the opponent's leg while stepping forward and to the left at approximately a 45-degree angle to close with the opponent.

Execute a finishing technique, such as a strike or a kick, to the opponent's exposed target areas.

Counter to a Front Kick (Right or Rear Leg). This counter is used when the opponent executes a front kick with his right leg. To execute the counter to a front kick, Marines—

Step forward and to the left at approximately a 45-degree angle, moving in to the outside of the opponent's striking leg.

Lower the left arm and block or deflect the opponent' s leg with the palm of the hand or the meaty portion of the forearm.

Hit and stick by leaving the left arm against the opponent's leg while stepping forward and to the right at approximately a 45-degree angle to close with the opponent.

CHAPTER 5

THROWS

This chapter describes all techniques for a right-handed person. However, all techniques can be executed from either side.

The Marine is depicted in camouflage utilities. The opponent is depicted without camouflage.

Marines use throwing techniques to maintain the tactical advantage and to throw the opponent to the ground during close combat. Throws apply the principles of balance, leverage, timing, and body position to upset an opponent's balance and to gain control by forcing the opponent to the ground. Throwing techniques are effective because they are size- and gender-neutral, and they rely on the momentum and power generated by the opponent rather than the strength or size of the Marine. Marines also execute a throw as a devastating attack against an opponent, possibly causing unconsciousness or broken limbs. When Marines execute throws, they must maintain balance and, simultaneously, prevent the opponent from countering a throw or escaping after being forced to the ground.

1. Turning Throw

Marines use a turning throw to take the opponent to the ground while they remain standing. A turning throw can also be executed from a stationary position. It is particularly effective if Marines and the opponent are wearing gear. To execute the turning throw, Marines-

Grasp the opponent's right wrist with the left hand.

Step forward with the right foot, place it against the outside of the opponent's right foot, and pivot so the back of the heel is next to the middle of the opponent's foot.

Hook the opponent's right arm with the right arm and pinch his arm between the biceps and forearm, touching the opponent with the body.

Pull the opponent's wrist downward, keeping it close to the body.

Pivot to the left on the ball of the foot and continue pulling downward on the opponent's wrist while rotating the wrist outward to off-balance him.

2. Hip Throw

Marines use a hip throw to take the opponent to the ground while they remain standing. A hip throw is particularly effective if the opponent is moving forward or pushing. Marines use the opponent' s forward momentum to execute the hip throw. To execute the hip throw, Marines-Grasp the opponent's right wrist with the left hand.

Step forward with the right foot and place it against the outside of the opponent's right foot.

The back of the heel should be next to the opponent' s foot.

Step back with the left foot and rotate on the ball of the foot. The back of the heel is next to the opponent's toe. The knees are bent.

Rotate at the waist and hook the right hand around the back of the opponent's body (anywhere from his waist to his head). The side and hip should be against the opponent.

Rotate the hip against the opponent. The hips must be lower than the opponent's.

Use the right hand to pull the opponent up on the hip to maximize contact.

Pull the opponent's arm across the body and, at the same time, lift the opponent off the ground slightly by bending at the waist, straightening the legs, and rotating the body to the left.

3. Leg Sweep

Marines use a leg sweep to take the opponent to the ground while they remain standing. A leg sweep is particularly effective if the opponent is already off-balanced and moving backward or pulling on the Marine. To execute the leg sweep, Marines—

Grab the opponent's right wrist with the left hand and grab the opponent's left shoulder with the right hand.

Note: Grab the opponent's clothing or gear if his wrist and shoulder cannot be grabbed.

Step forward with the left foot and place it on the outside of the opponent's right foot.

Pull the opponent's wrist downward, close to the body, and push his shoulder backward to off-balance him.

Raise the right knee no higher than the waist.

Kick the foot past the opponent's right leg.

Use the heel of the boot to make contact with the opponent's calf (anywhere from the top of the calf down to the Achilles tendon or on the inside of the calf).

Sweep the opponent to the ground.

CHAPTER 6

CHOKES AND HOLDS

This chapter describes all techniques for a right-handed person. However, all techniques can be executed from either side.

The Marine is depicted in camouflage utilities. The opponent is depicted without camouflage.

WARNING

During training, never execute a choke at full force or full speed and never hold a choke for more than 5 seconds.

When Marines correctly perform a choke, they render an opponent unconscious in as little as 8 to 13 seconds. Chokes are easily performed regardless of size or gender. Marines must know how to apply chokes and how to counter a choke or a hold executed by an opponent.

1. Types of Chokes

There are two types of chokes: an air choke and a blood choke. An air choke closes off the airway to the lungs, thereby preventing oxygen from reaching the heart. A blood choke cuts off the blood flow to the brain. Both types can result in unconsciousness and eventual death for an opponent.

Air Choke

An air choke is performed on the opponent's windpipe (or trachea), cutting off the air to the lungs and heart. If Marines execute the air choke properly, the opponent loses consciousness within 2 to 3 minutes. Due to the length of time it takes to immobilize the opponent, air chokes are not recommended.

Blood Choke

A blood choke is performed on the opponent's carotid artery, which carries oxygen-enriched blood from the heart to the brain. The carotid artery is located on both sides of the neck. If Marines execute a blood choke properly, the opponent will lose consciousness within 8 to 13 seconds. The blood choke is the preferred choke because its intended effect (i.e., the opponent losing consciousness) can be executed quickly, ending the fight.

2. Chokes

Front Choke

Marines execute a front choke when they are facing the opponent. Marines use the opponent's lapels or collar to execute a front choke. To execute the front choke, Marines-

Grab the opponent's right lapel with the right hand, making certain that the knuckles or the back of the hand are against the opponent's carotid artery on the right side of his neck.

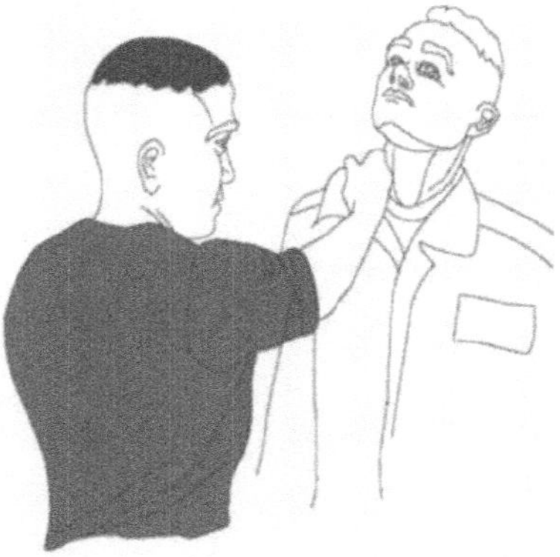

Keep the right hand against the opponent's neck, reach under the opponent's right arm with the left hand, grab the opponent's left lapel, and form an X with the wrists.

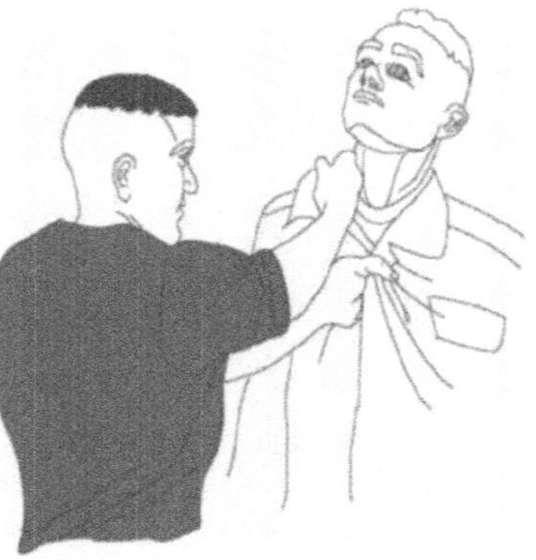

Press the right hand against the opponent's carotid artery.

Pull his lapel to the left with the left hand.

Side Choke

Marines execute a side choke when they are facing the opponent. The side choke is particularly effective when deflecting a punch thrown by an opponent. To execute the side choke, Marines—

Use the left hand to parry the opponent's arm inboard (to the inside of the opponent's reach).

Bring the right arm underneath the opponent's arm and up around the front of his neck.

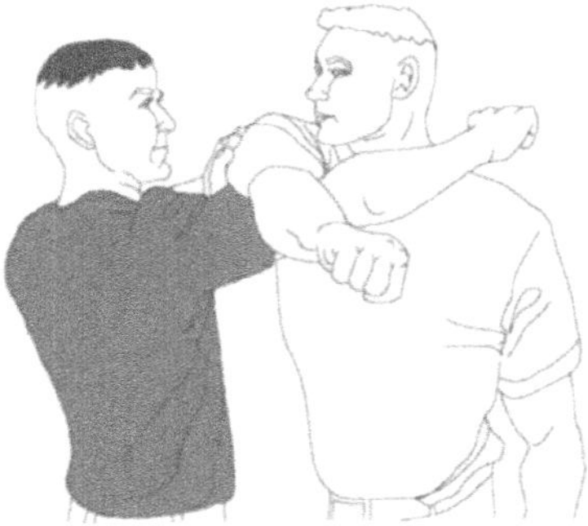

Extend the fingers, place the back of the forearm against the opponent's neck just below his ear, and press the carotid artery on his neck.

Note: If Marines are unable to place their fingers against the opponent's neck, they may use the back of the thumb or wrist.

Reach, with the left hand, around the back of the opponent's neck and clasp the hands together.

Pull the opponent toward the chest by pulling the clasped hands toward the chest.

Exert pressure on the side of the opponent's neck with the forearm.

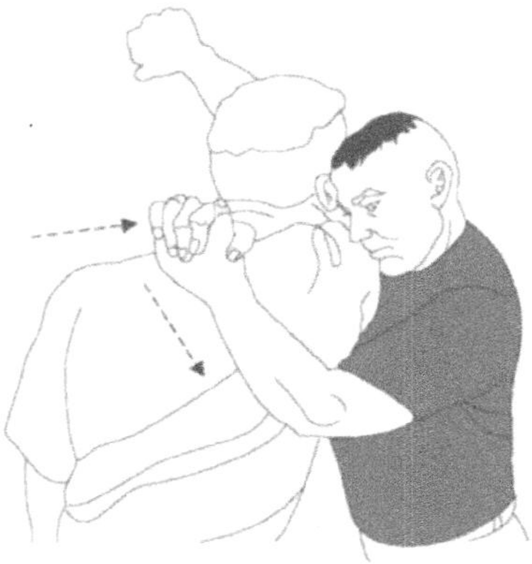

Rear Choke

Marines execute a rear choke when they are behind the opponent, the opponent is on the ground, or when they are taking the opponent to the ground. To execute the rear choke, Marines—

Reach, with the right arm, over the opponent's right shoulder and hook the bend of the arm around his neck.

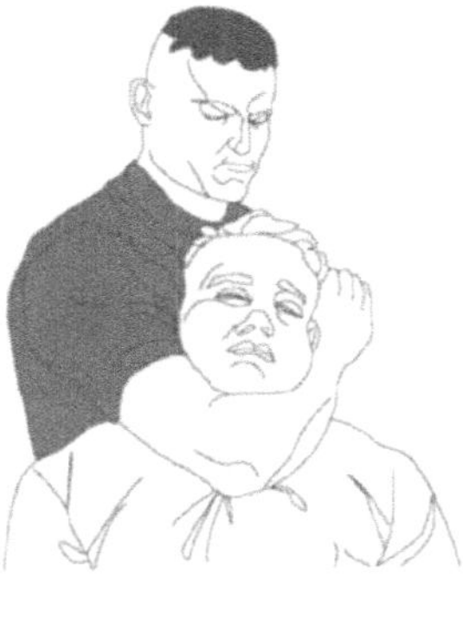

Clasp both hands together.

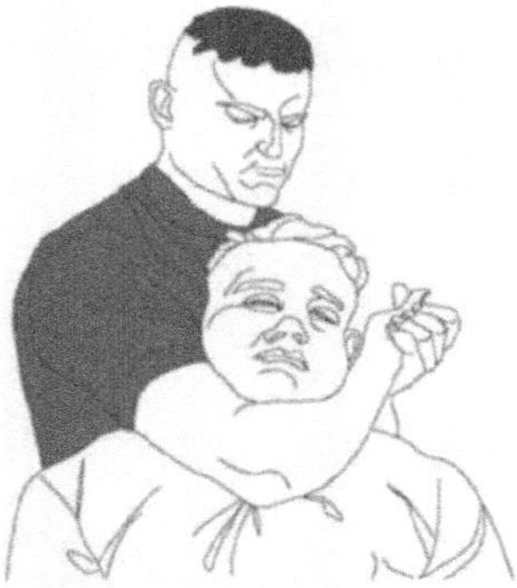

Exert pressure with the biceps and forearms on both sides of the opponent's neck on his carotid arteries.

Maintain pressure with the biceps and forearms on both sides of the neck and draw the opponent closer by drawing the right arm in.

Figure-Four Choke

A variation of the rear choke is the figure-four choke. The figure-four choke allows Marines to gain more leverage than the rear choke. If the rear choke cannot be secured, the figure-four variation is employed to increase the pressure of the choke on the opponent. To execute the figure-four variation of the rear choke, Marines—

Apply a rear choke. The Marine's body should be against the opponent's body.

Grasp the left biceps with the right hand and place the left hand against the back of the opponent's head.

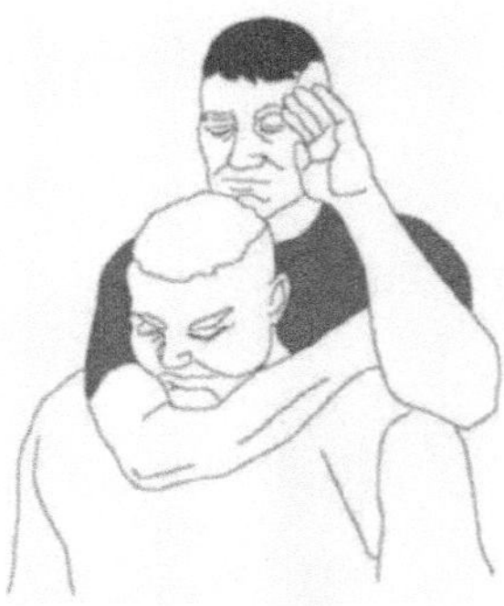

Push the opponent's head forward and down with the left hand.

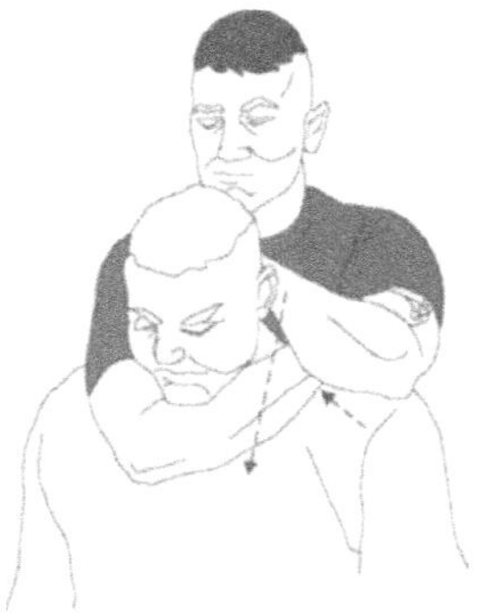

Draw the right arm in, maintaining pressure with the bicep and forearm on both sides of the opponent's neck.

3. Counters to Chokes and Holds

During a close combat situation, an opponent may apply a choke or hold on a Marine. If the opponent correctly applies a choke, a Marine quickly loses consciousness. If a choke is not executed properly, it often results in a hold, typically a bear hug or a headlock. A hold allows the opponent to control a Marine and removes the Marine's ability to attack. It is important for Marines to extract themselves from chokes and holds, regain the tactical advantage, and counter with strikes.

Although a choke causes unconsciousness in 8 to 13 seconds for a blood choke and 2 to 3 minutes for an air choke, the first movement in any counter to a choke is to clear the airway. Marines use softening techniques to loosen an opponent's grip and to clear their airway. Softening techniques are particularly effective if Marines lack the physical strength of their opponent. These techniques include groin strikes, eye gouges, foot stomps, etc. Softening techniques are not offensive; rather, they are used to loosen an opponent's hold.

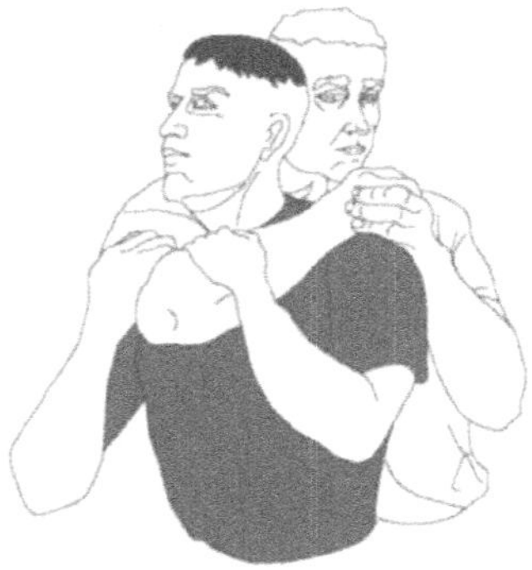

The second movement is to tuck the chin. Once the airway is clear, Marines tuck their chins to prevent the opponent from reapplying the choke.

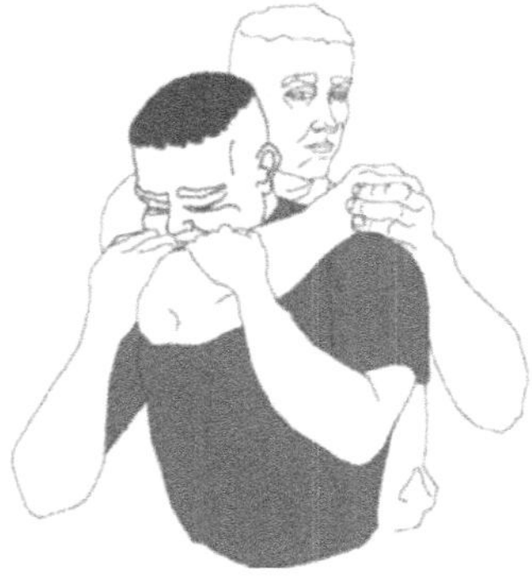

Counter to a Front Choke

Marines use a counter to a front choke when the opponent approaches from the front and uses both hands to choke a Marine around the throat. To execute the counter to the front choke, Marines—

Grasp the opponent's right forearm (where the elbow bends) with the left hand and apply downward pressure on the opponent's radial nerve with the fingers.

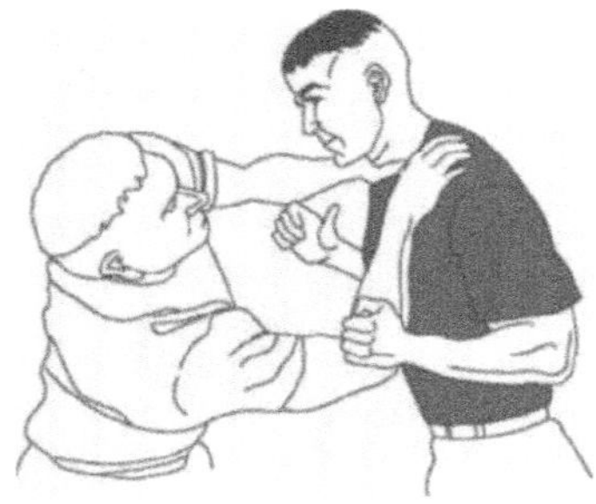

Execute a chin jab to the opponent's chin with the right hand. To generate power into the strike, bring the left foot to the outside of the opponent's right foot and rotate the hips into the strike.

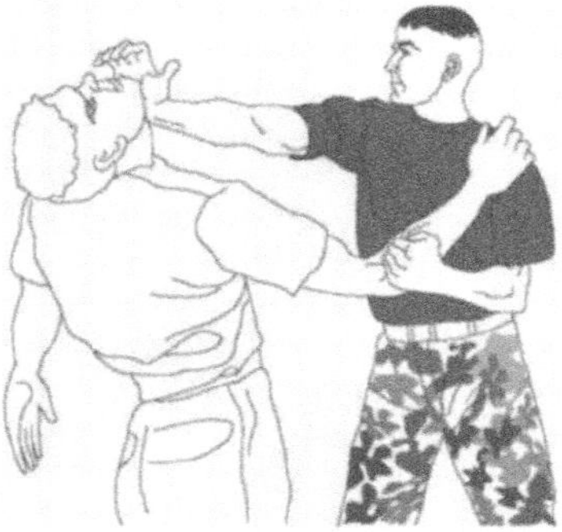

Counter to a Rear Choke

Marines execute a counter to a rear choke when the opponent approaches from the rear and puts his right arm around a Marine's throat. To execute the counter to the rear choke, Marines-

Grasp the opponent's forearm (at the radial nerve) and bicep with both hands and pull down just enough to clear the airway. Once the airway is clear, tuck the chin to protect the airway and to prevent the opponent from reapplying the choke.

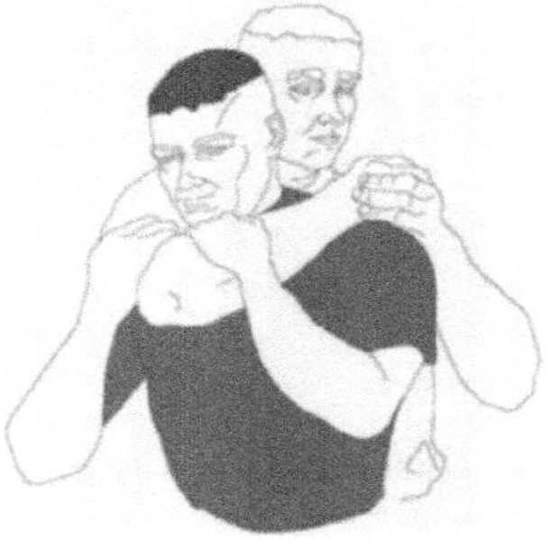

Step behind the opponent's right leg with the left foot, keeping both legs bent (almost in a squatting position).

Strike and drive the left elbow into the opponent's torso while rotating the hips and pivoting to the left, throwing the opponent back and to the ground.

Counter to a Front Headlock

Marines use a counter to a front headlock when the opponent approaches from the front and puts his right arm around the Marine's neck, bends the Marine forward, and locks the Marine's head against his hip. To execute the counter to a front headlock, Marines—

Grasp the opponent's wrist and forearm with both hands and pull down to clear the airway. Maintain control of the opponent's wrist throughout the move. Once the airway is clear, tuck the chin to protect the airway and to prevent the opponent from reapplying the choke.

Move the right hand and arm across the opponent' s torso.

Step forward and to the left with the left foot at a 45-degree angle.

Execute a sweep with the right foot against the opponent's right leg. At the same time, push against the opponent's chest with the right arm and shoulder to generate power in the sweep.

Counter to a Rear Headlock

Marines use a counter to a rear headlock when the opponent approaches from the rear and puts his right arm around the Marine's neck, bends the Marine forward, and locks the Marine's head against his hip. To execute the counter to the rear headlock, Marines-

Grasp the opponent's wrist and forearm with the right hand and pull down to clear the airway. Once the airway is clear, tuck the chin to protect the airway and to prevent the opponent from re-applying the choke.

Reach over the opponent's right shoulder with the left arm.

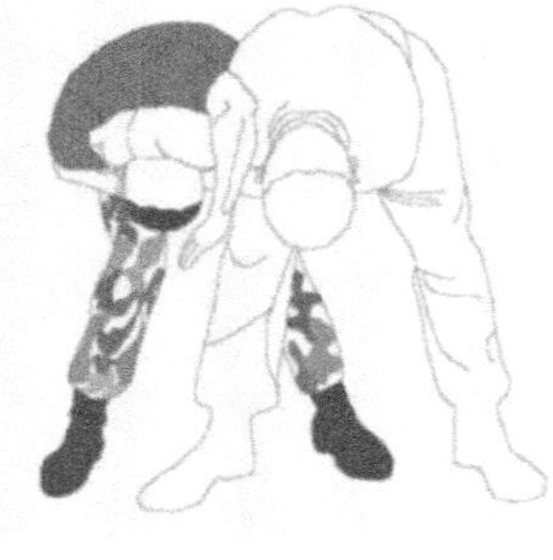

Grab any part of the opponent's face (chin, nose, eyes) and pull back while rising to a standing position.

Execute, with the right hand, a hammer fist strike to the opponent's exposed throat.

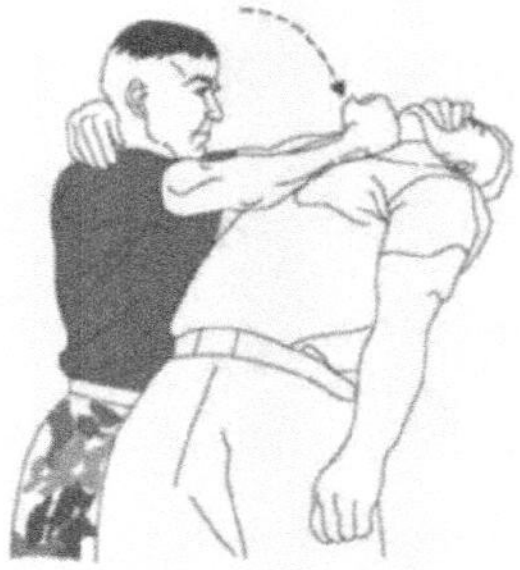

Counter to a Front Bear Hug

Marines execute a counter to a front bear hug when the opponent approaches from the front and puts both of his arms around the Marine's body, trapping the Marine's arms to the sides. To execute the counter to a front bear hug, Marines—

Step forward and to the left with the left foot at a 45-degree angle to the outside of the opponent's right leg, keeping the left leg bent.

Grasp the opponent's torso or arms to gain balance and to assist in throwing the opponent. It may be helpful to hook the opponent's right arm with the left arm.

Drive the right arm and shoulder forward and, at the same time, bring the right leg forward and sweep the opponent's right leg, bringing him to the ground.

Counter to a Rear Bear Hug

Mar:nes use a counter to a rear bear hug when the opponent approaches from the rear and puts both of his arms around the Marine's body, trapping the Marine's arms to the sides. To execute the counter to a rear bear hug, Marines—

Step behind the opponent's right leg with the left foot, keeping both legs bent (almost in a squatting position). The left side of the body should be against the opponent's.

Pivot the hip, turning the body to the left and throwing the opponent back over the bent leg.

CHAPTER 7

GROUND FIGHTING

This chapter describes all techniques for a right-handed person. However, all techniques can be executed from either side.

In drawings, the Marine is depicted in woodland camouflage utilities; the opponent is depicted witlwut camouflage. In photographs, the Marine is depicted in woodland camouflage utilities; the opponent is depicted in desert camouflage utilities.

Marines should avoid being on the ground during a close combat situation because the battlefield may be covered with debris and there is an increased risk of injury. However, many close combat situations involve fighting on the ground. The priority in a ground fight is for Marines to get back on their feet as quickly as possible. In any ground fighting scenario, Marines will usually end up in one of four positions with the opponent. The offensive positions, in which Marines have a tactical advantage, are the guard and mount. The defensive positions, which are used as counters when the opponent has the tactical advantage, are the counter to the guard and the counter to the mount. Marines can also employ chokes during ground fighting to quickly end a fight.

1. Offensive Ground Fighting

Guard Position

Marines execute a guard position when the opponent is on top and the legs are wrapped around the opponent's legs. If the opponent is on top, he may try to choke the Marine, but the Marine still has the tactical advantage because the Marine is in a position to control the situation. To execute the guard position technique, Marines—

Trap the opponent's hands on the chest by crossing the hands on the chest so the forearms are resting on the opponent's forearms. Apply pressure with the elbows or forearms.

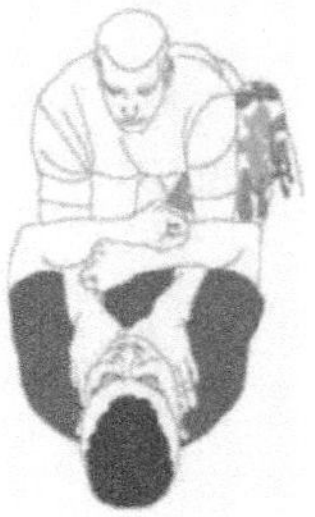

Strike the outside of the opponent's thigh with the cutting edge of the right heel. This causes the opponent to jerk to that side.

Move the head quickly to the left and swivel the hips to the right. At the same time, bring up both of the legs. Both legs are on the right side of the opponent's body.

Bring the right leg down, hooking the opponent's neck and head. Exert downward pressure to roll him over on his back. Grasp and maintain control of the opponent's left arm. Upon completion of the move, the Marine is sitting up with legs bent over the opponent while maintaining control of his arm.

Keep legs and knees bent. Maintain pressure against the opponent's neck with the back of the right foot and against his side with the left foot underneath his armpit. Squeeze the knees together, locking the opponent's arm.

Pull the opponent's arm straight up and fall back sharply, pulling his arm to the side in the direction of his little finger.

Return to standing.

Mount Position

Marines execute the mount if the opponent is lying on his back on the ground and the Marine is on top with legs wrapped around the opponent's body. This position is an offensive position because Marines are in a better position to control the opponent and to execute ground fighting techniques. To execute the mount position technique, Marines—

Grab the opponent's wrists or forearms with the hands. Hold them tightly against the torso.

Maintain control of the opponent's arms with the left arm and apply pressure to the opponent's brachial plexus (tie in) with the right hand.

Bring the left foot close to the opponent's right armpit and swing the right leg across the opponent' s head.

Keep the legs and knees bent. Maintain pressure against the opponent's neck with the back of the right foot and against his side with the left foot which is underneath his armpit. Squeeze the knees together, locking the opponent's arm.

Pull the opponent's arm straight up and fall back sharply, pulling his arm to the side in the direction of his little finger.

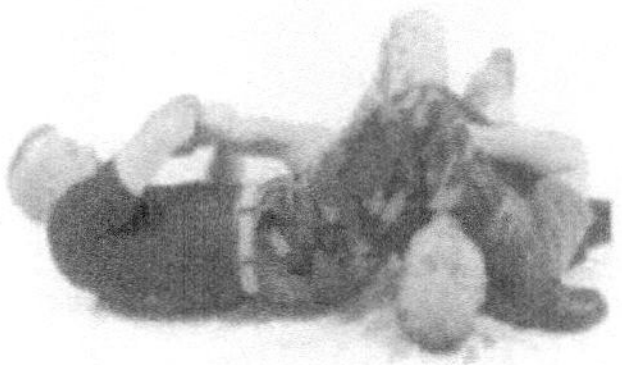

Return to standing.

2. Defensive Ground Fighting

Counter to the Guard

Marines use this technique if the opponent is lying on his back on the ground and the Marine is kneeling on the ground between the opponent's legs. To execute the counter to the guard position, Marines—

Strike the opponent's femoral nerve, located on the inside of the thigh, with the elbows. This forces the opponent to separate his legs.

Strike the opponent's groin with the right fist.

Hook the left arm underneath the opponent's right knee from the inside. Throw the opponent's leg over the head.

Duck the head quickly and move to the left.

Return to standing.

Counter to the Mount

Marines use this technique when they are lying on their back on the ground, the opponent is mounted on top of the Marine, and the opponent's legs are wrapped around the Marine. The opponent has the tactical advantage. To execute the counter to the mount position, Marines—

Grab the opponent's gear or clothing on his upper torso and pull him down close.

Use the right arm to hook the opponent's left arm, from the inside around the outside, above his elbow. With the right foot, hook the opponent's left leg or ankle.

Draw the elbow in to bend the opponent's elbow, bringing him down close. The arm must be hooked above the opponent's elbow in order to bend it.

Strike the opponent's side with the left hand.

Push the opponent over and roll him off to the right side.

Maintain control of the opponent's hooked arm and move to a kneeling position.

Move to a standing position while maintaining pressure on the opponent's arm and using the knee to apply pressure against the opponent's elbow.

3. Ground Fighting Chokes

WARNING

During training, never execute a choke at full force or full speed. Never hold a choke for more than 5 seconds.

The priority in ground fighting is for Marines to get back on their feet as quickly as possible. Sometimes, Marines can quickly end a ground fight by executing a choke on the opponent. When performed correctly, a choke can render an opponent unconscious in as little as 8 to 13 seconds. Chokes are easily performed regardless of size or gender. The chokes performed during ground fighting are the same as those performed while standing (see chap. 6).

Ground Fighting Front Choke

The ground fighting front choke is a blood choke performed most effectively from the mount position. The front choke employs the opponent's lapels or collar to execute the choke. To execute the ground fighting front choke, Marines—

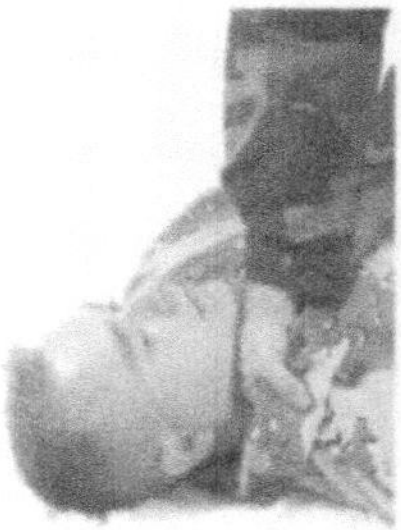

Grab the opponent's right lapel with the right hand, making certain that the knuckles or the back of the hand are against the carotid artery on the right side of the opponent's neck.

Keep the right hand pressed against the opponent's carotid artery, reach under the right arm with the left hand, and grab the opponent's left lapel, forming an X with the wrists.

Keep the right hand pressed against the opponent's carotid artery and pull the opponent's left lapel to the left with the left hand.

Ground Fighting Side Choke

The ground fighting side choke is a blood choke performed from the mount position. The ground fighting side choke is particularly effective when the opponent raises his arms and places them on the Marine's chest or throat. To execute the ground fighting side choke, Marines-

Use the left hand to parry the opponent's right arm inboard (to the inside of the opponent's reach). Bring the right arm underneath the opponent's arm and up around the front of his neck.

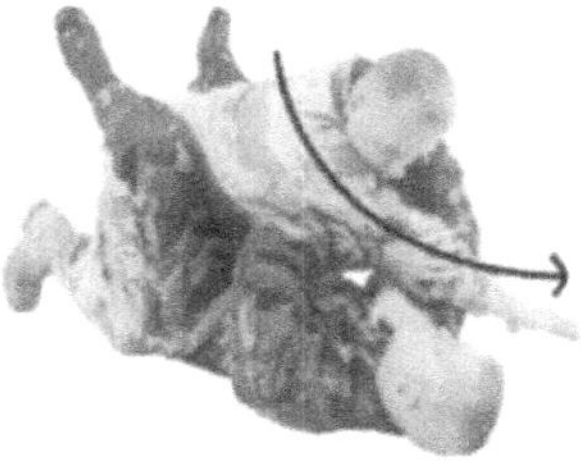

Keep the fingers extended, place the back of the forearm against the opponent's neck just below his ear and press the carotid artery on his neck. Depending on the Marine's or the opponent's body configuration, the back of the thumb or wrist may press against the opponent's neck.

Use the left hand to reach around the back of the opponent's neck and clasp the hands together.

Pull the opponent toward the chest. Use the forearm to exert pressure on the side of his neck. This is done by pulling the clasped hands toward the chest.

Ground Fighting Rear Choke

The ground fighting rear choke is a blood choke performed when Marines are behind the opponent. To execute the ground fighting rear choke, Marines—

Wrap the legs around the opponent with the left foot against the inside of his left thigh and the right foot against the inside of his right thigh.

Use the right arm to reach over the opponent's right shoulder and hook the bend of the arm around his neck.

Use the left hand to clasp both hands together.

Exert pressure with the biceps and forearms on both sides of the opponent's neck. This applies pressure to the opponent's carotid arteries. While maintaining pressure with the biceps and forearms, draw the opponent closer by drawing the right arm in. To increase the effectiveness of the choke, lean backward by arching the back and pulling the opponent back. At the same time, push the feet against the opponent's thighs.

Figure-Four Choke

A variation of the ground fighting rear choke is the figure-four choke. The figure-four choke allows Marines to gain more leverage on the rear choke. If the rear choke cannot be secured, the figure-four variation may be applied to increase the pressure of the choke on the opponent. To execute the figure-four variation of the rear choke, Marines—

Apply the ground fighting rear choke. The chest should be against the opponent's back.

Grasp the left biceps with the right hand and place the left hand against the back of the opponent's head.

Use the left hand to push the opponent's head forward and down.

Draw the right arm in while maintaining pressure with the biceps and forearm on both sides of the opponent's neck.

Increase the effectiveness of the choke by arching backward and pushing the feet against the opponent's thighs. Continue to pull the opponent back with the right arm while exerting pressure forward with the left hand against the opponent's head.

CHAPTER 8

NONLETHAL TECHNIQUES

This chapter describes all techniques for a riglit-handed person. However, all techniques can be executed from either side.

In drawings, the Marine is depicted in woodland camouflage utilities; the opponent is depicted without camouflage. In photographs, the Marine is depicted in woodland camouflage utilities; the opponent is depicted in desert camouflage utilities.

The Marine Corps' involvement in military operations other than war—e.g., humanitarian, peacekeeping, or evacuation missions—has greatly increased. These missions require skills that span the spectrum of conflict and support operations within a continuum of force. But the Corps' day-to-day existence also demands a responsible use of force. Nonlethal techniques are among the skills Marines use to apply a responsible use of force.

1. Unarmed Restraints and Manipulation

Marines operate within a continuum of force, particularly in support of peacekeeping or humanitarian types of missions. In these situations, Marines must act responsibly to handle a situation without resorting to deadly force. Unarmed restraints and manipulation techniques including joint manipulation, come-alongs, and takedowns can be used to control a subject without resorting to deadly force. Marines must train to become proficient in nonlethal techniques and to respond in a responsible manner. These techniques are referred to as compliance techniques, and they are applied in the third level in the continuum of force.

WARNING

During training, never apply the techniques for unarmed restraints and manipulation at full force or full speed. Use a slow and steady pressure to avoid injury.

Compliance Techniques

Compliance techniques are unarmed restraint and manipulation techniques used to physically force a subject or opponent to comply. Compliance can be achieved through the close combat techniques of-

Pain compliance using joint manipulation and pressure points. (Pain compliance is the initiation of pain to get compliance on the part of the subject.)

Come-along holds.

Principles of Joint Manipulation

Joint manipulation is used to initiate pain compliance and gain control of a subject. It involves the application of pressure on the joints (elbow, wrist, shoulder, knee, ankle, and fingers). Pressure is applied in two ways:

In the direction in which the joint will not bend. For example, joints such as the knees

and elbows only bend in one direction and when pressure is applied in the opposite direction, pain compliance can be achieved. Beyond the point where the joint stops naturally in its range of movement (i.e., it no longer bends).

Since each joint has a breaking point, Marines should apply slow steady pressure only until pain compliance is reached. Continued pressure will break the joint and may escalate the violence of the situation.

Joint manipulation also uses the principle of off-balancing. A subject can be better controlled when he is knocked off balance.

Wristlocks

A wristlock is a joint manipulation that can be applied in a number of ways to achieve pain compliance. The wrist rotates in a number of directions and will bend in a single direction until its movement stops naturally. In a wristlock, pressure is exerted beyond that point by bending or twisting the joint. A wristlock is executed when an opponent tries to grab Marines or is successful in grabbing Marines or their equipment. A wristlock can also be performed by Marines if they wish to initiate control of an opponent.

Basic Wristlock. A basic wristlock is executed when Marines grab the opponent's left hand with the right hand. To execute the basic wristlock, Marines—

Use the right hand to grab the opponent's left hand by placing the thumb on the back of the opponent's hand so that the Marine's knuckles are facing to the left.

Hook the fingers across the fleshy part of the opponent's palm below the thumb. The fingers are used to anchor the hand so leverage can be applied to twist and bend the joint.

Exert downward pressure with the thumb to bend the opponent's joint. Rotate the opponent's hand to the right to twist the joint.

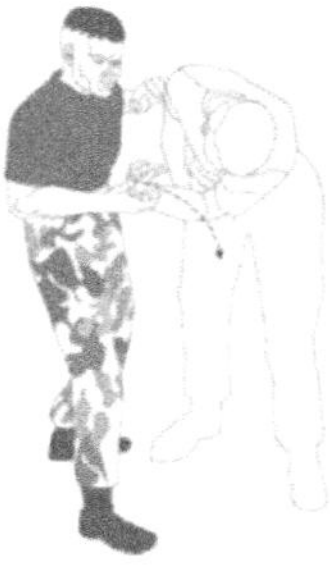

Use the left hand to further control the opponent.

Step in to the opponent to keep the opponent's hand in close to the body to control him and provide more leverage on the wristlock.

Note: When executing the basic wristlock with the left hand, the Marine grabs the opponent's hand so that the Marine's knuckles are facing to the right, and then rotates and twists the opponent's hand to the left.

Reverse Wristlock. A reverse wristlock is executed when Marines grab the opponent's right hand with the right hand. To execute the reverse wristlock, Marines—

Place the right palm on the back of the opponent's right hand and wrap the fingers across the fleshy part of his palm below his little finger.

Twist the opponent's hand to the right while stepping in to place his hand against the chest. Apply downward pressure on the opponent's hand against the chest. Leave the opponent's hand on the chest to fully control the subject and to gain leverage.

Lean forward to use body weight to add additional pressure to the joint.

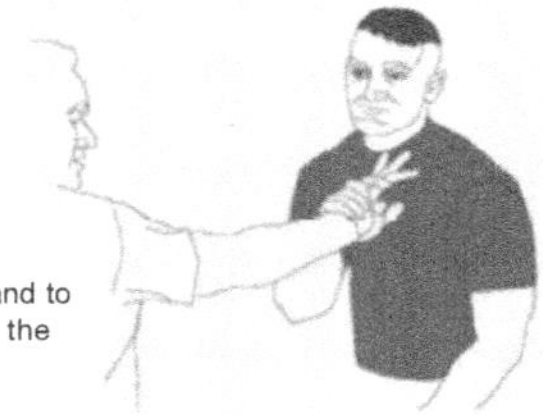

Use the left hand to further control the opponent.

Two-Handed Wristlock. Both hands can be used in the wristlock to maximize the leverage and pressure needed to bend and twist the joint. To execute the two-handed wristlock, Marines—

Place both thumbs on the back of the opponent's hand, thumbs crossed.

Hook the fingers of both hands around the fleshy part of the opponent's palm on both sides of his hand.

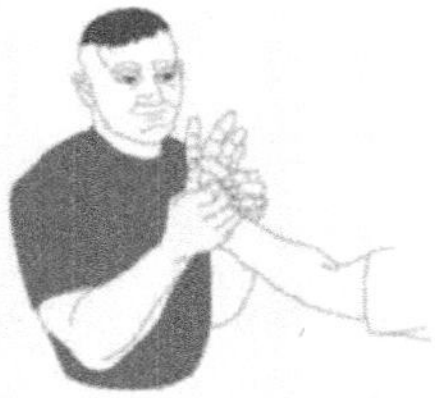

Step into the opponent and apply pressure downward on the back of his hand to bend the joint and rotate his wrist away from the body to twist the joint.

Enhanced Pain Compliance on Wristlock. Enhanced pain compliance techniques are applied in the third and fourth levels in the continuum of force. Additional pain be applied to a wristlock by-

Adding downward pressure to the elbow with the other hand or elbow by using the fingers to pull in on the opponent's radial nerve located on the inside of the forearm. When pressure is added to the opponent's radial nerve, his direction can be controlled.

Applying pressure against the opponent's finger joint to bend it in a direction it cannot bend (i.e., splitting the fingers).

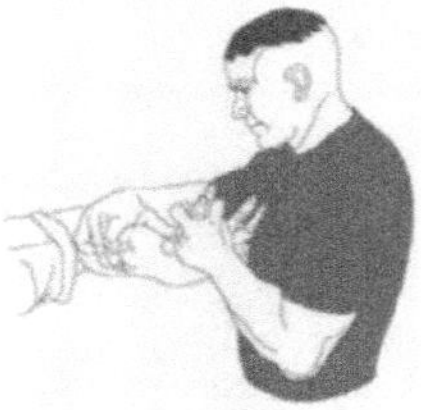

Come-Along Holds

Marines use a come-along hold to control and move an opponent.

Escort Position. A common come-along hold is the escort position. To execute the escort position, Marines—

Face the opponent. Use the left foot to step forward at a 45-degree angle. Turn to face the right side of the opponent.

Use the right hand to firmly grasp the opponent's right wrist. With the left hand, firmly grasp the opponent's right triceps.

Position the opponent's controlled arm diagonally across the torso, keeping his wrist against the right hip. The Marine should be standing to the right of and behind the opponent.

Note: This technique works well when escorting an opponent on either the right or left side. Take caution when escorting an opponent by ensuring his controlled hand is not in a position to grab the holstered weapon. The preferred escort position is from the Marine's left side, so that the opponent is kept further away from the weapon.

Wristlock Come-Along. To execute the wristlock come-along, Marines-

Use the left hand to execute a basic wristlock. Incorporate the right hand in a two-handed wristlock for more control.

Maintain pressure on the opponent's wrist with the right hand, step forward, and pivot around to stand next to the opponent.

Release the left hand, quickly reach under the opponent's arm from behind, and grab his hand.

Use the left hand and apply downward pressure on the opponent's wrist.

Controlling Technique. The following controlling technique is used when an opponent grabs the Marine's wrist. To execute the technique, Marines—

Trap the opponent's hand with the palm of the other hand.

Rotate the opponent's trapped hand up and on his forearm while maintaining downward pressure on his trapped hand.

Apply downward pressure with both hands until the opponent is taken to the ground.

Armbars

An armbar is a joint manipulation in which pressure is applied on a locked elbow, just above the joint, in the direction the joint will not bend. An armbar has to be locked in quickly, but still requires a slow, steady pressure to gain compliance.

Basic Armbar. To execute a basic armbar, Marines-

Use the right hand to grab the opponent's right wrist.

Bring the left hand down on or above the opponent's elbow joint. To gain additional leverage, pivot to face the opponent.

Use the left hand to apply downward pressure on or above the opponent's elbow joint while pulling up on his wrist.

Armbar from a Wristlock. To execute an armbar from a wristlock, Marines-

Use the right hand to grab the opponent's right hand and execute a reverse wristlock.

Bring the left hand down on or above the opponent's elbow joint.

Use the left hand to apply downward pressure on or above the opponent's elbow joint while pulling up on his wrist.

Takedowns

A takedown is used to bring an opponent to the ground to further control him.

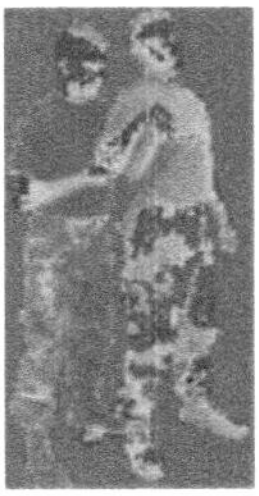

Takedown From a Wristlock Come-Along. To take the opponent to the ground from a wristlock come-along, Marines—

Use the right foot to push down on the opponent's calf or Achilles tendon.

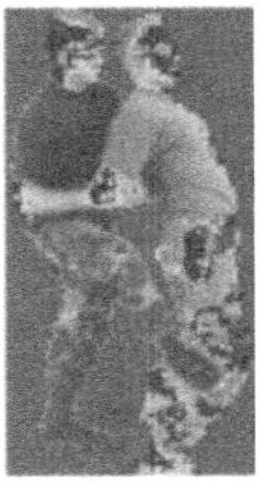

Maintain control of the opponent's wrist and elbow and apply a slow, steady pressure to bring him to the ground.

Armbar to a Takedown. This technique is used to take a noncompliant opponent to the ground from an armbar. To execute the armbar takedown, Marines—

Use the right hand to execute a reverse wristlock.

Bring the left hand or forearm down on or above the opponent's elbow joint.

Pivot so the back is facing the opponent and, at the same time, lift the left elbow and slide the body so it is against the opponent, placing the armpit high above the opponent's elbow joint.

Lean back, placing the body weight on the opponent's arm until he complies or is taken to the ground.

Note: This technique may break the opponent's arm. Therefore, this technique should not be employed if the objective is a nonlethal takedown.

Wristlock Takedown. This technique is used to take a noncompliant opponent to the ground from a basic wristlock and to put him in a position where he can be handcuffed, if necessary. To execute the takedown, Marines-

Use the right hand to execute a basic wristlock.

Incorporate the left hand in a two-handed wristlock.

Apply downward pressure on the wristlock, pivot on the ball of the right foot, and quickly turn to the right to take the opponent to the ground.

Continue to apply pressure on the wristjoint as the opponent lands on his back with his arm straight in the air.

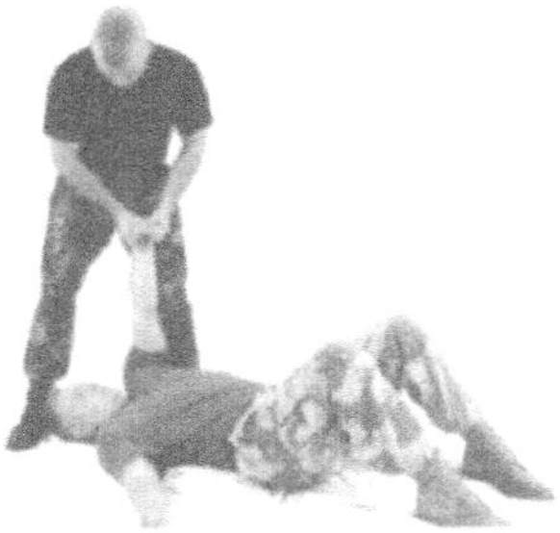

Slide the left foot under the opponent's back, underneath his armpit.

Apply pressure with the knee against the opponent's triceps while pulling back on his arm.

Continue applying pressure with the knee on the opponent's arm. Pivot and step around the opponent's arm to roll him on his stomach.

Kneel down with one knee on the opponent's back. The other knee is placed on the opponent's neck and shoulder on either side of his arm. Apply inward pressure with the knees to lock his arm in place.

Tell the opponent to put his other hand in the middle of his back. Bring the opponent's controlled hand to the center of his back.

Escort Position Takedown. This technique can be used to control a noncompliant opponent from the escort position. To execute the escort position takedown, Marines—

Lock the opponent's arm straight across the body while rotating his wrist away from the body.

Use the left hand or forearm and apply downward pressure above the opponent's elbow where the triceps meet.

Step back with the right foot and, keeping the opponent's hand controlled against the hip, pivot to the right while continuing to apply downward pressure on his arm to bring him to the ground.

Note: This technique works well when escorting a subject on the Marine's right or left side. When taking down a subject from the right side, step back and pivot to the left.

2. Nonlethal Baton

A baton or nightstick can be an effective compliance tool when used correctly. Batons can be used defensively (blocking), offensively (striking), and as a restraining device when needed. In the fourth level of the continuum of force (assaultive [bodily harm]), defensive tactics include baton or nightstick blocks and blows. Blows to the head or other bony parts of the body are considered deadly force. When deadly force is not authorized, Marines must be able to employ blocks, strikes, and restraints with the baton with the minimum of force.

Grips

One-Handed Grip. To execute the one-handed grip, Marines—

Use the right hand to grasp the lower end of the baton, about 2 inches from the end.

Wrap the thumb and index finger around the baton so they are touching one another. The grip on the baton should be firm, but natural.

Two-Handed Grip. To execute the two-handed grip, Marines-

Use the right hand to grasp the lower end of the baton, about 2 inches from the end. Wrap the thumb and index finger around the baton so they touch one another.

Use the left hand to grasp the upper end of the baton, palm down, about 2 inches from the end. The hands should be approximately 10 to 12 inches apart.

Stance and Method of Carry

The basic warrior stance serves as the foundation for initiating nonlethal baton techniques. The method of cany provides effective defensive positions with a wide range of options to control a combative opponent.

One-Handed Carry. To execute the one-handed carry, Marines—

Grip the baton using the one-handed grip.

Elevate the baton, with the gripping hand at a level between the belt and shoulder.

Keep the left hand in the position of the basic warrior stance.

Two-Handed Carry. This carry is effective for blocks. To execute the two-handed carry, Marines-

Grip the baton using the two-handed grip.

Elevate the baton, with the left hand higher than the right hand.

Orient the weapon toward the subject.

Movement

In a nonlethal confrontation, movement may be made to create distance between Marines and the opponent or to close the gap to control him. When facing an opponent, Marines move in a 45-degree angle to either side of the opponent. Moving at a 45-degree angle is the best way to both avoid an opponent's strike and to put Marines in the best position to control the opponent.

Blocking Techniques

One-handed blocks are used when carrying the baton in the one-handed carry. The same one-handed blocks used in combative stick techniques (see page 3-6) apply in nonlethal baton engagements. The blocks include blocks for a vertical strike, a forward strike, and a reverse strike. Because the baton is often carried with two hands, there are also two-handed blocks that are effectively used from this carry. Two-handed blocks are discussed in the following subparagraphs.

High Block. Marines execute a high block to deter a downward vertical attack directed at the head and shoulders. To execute the high block, Marines—

Raise the baton up to a level even with or above the head. The baton should be in a horizontal position to block the blow. The fingers of the left hand should be open and behind the baton.

Place the baton perpendicular to the opponent's striking surface to absorb the impact of the blow.

Bend the elbows slightly to help absorb the impact of the blow. The arms should give with the strike of the blow.

Low Block. Marines execute a low block to deter an upward vertical attack directed at the abdomen, groin, or torso. The opponent's blow can be delivered by a foot, knee, or fist. To execute the low block, Marines—

Lower the baton to a level even with or below the groin. The baton should be in a horizontal position to block the blow. The fingers of the left hand should be open and behind the baton.

Place the baton perpendicular to the opponent's striking surface to absorb the impact of the blow.

Bend the elbows slightly to help absorb the impact of the blow. The arms should give with the strike of the blow.

Right Block. Marines execute a right block to deter an attack directed at the head, neck, flank, or hip. The opponent's blow can be delivered by a foot, knee, fist, or elbow. To execute the right block, Marines-

Thrust the baton in a vertical position to the right side.

Pivot to the right by stepping with the left foot and pivoting off the ball of the right foot. Rotate the hips and shoulders into the direction of the block. The fingers of the left hand should be open and behind the baton.

Place the baton perpendicular to the opponent's striking surface to absorb the impact of the blow.

Bend the elbows slightly to help absorb the impact of the blow. The arms should give with the strike of the blow.

Left Block. Marines execute a left block to deter an attack directed at the head, neck, flank, or hip. The opponent's blow can be delivered by a foot, knee, fist, or elbow. To execute the left block, Marines—

Thrust the baton in a vertical position to the left side.

Pivot to the left by stepping with the right foot and pivoting off the ball of the left foot. Rotate the hips and shoulders into the direction of the block. The fingers of the left hand should be open and behind the baton.

Place the baton perpendicular to the opponent's striking surface to absorb the impact of the blow.

Bend the elbows slightly to help absorb the impact of the blow. The arms should give with the strike of the blow.

Middle Block. Marines execute either a left or right block to deter an attack directed at the face, throat, chest, or abdomen. To execute the middle block, Marines—

Thrust the baton in a vertical position straight out in front of the body. The fingers of the left hand should be open and behind the baton.

Place the baton perpendicular to the opponent's striking surface to absorb the impact of the blow. The baton should be held with the left hand forward of the right.

Bend the elbows slightly to help absorb the impact of the blow. The arms should give with the strike of the blow.

Restraining Technique

The strong-side armlock is used to restrain an opponent who is not compliant. To execute the strong-side armlock, Marines—

Use the right hand to run the baton up under the opponent's left armpit, parallel to the ground.

Use the right foot to step forward at a 45-degree angle to the left side of the opponent. The baton is across his forearm.

Use the right hand to drive the baton forward and up so the action bends the opponent's arm behind his back. At the same time, continue moving around him to get behind him.

Place the baton on the opponent's forearm with the thumb and/or fingers and apply pressure to his forearm. At the same time, grasp the other end of the baton with the left hand.

Pull up on the low end of the baton with the right hand. At the same time, push down on the top end of the baton with the left forearm, reaching around with the left hand to grasp the opponent's biceps or shoulder.

Continue exerting downward pressure with the left forearm while pulling back on the opponent's biceps with the left hand. This places the opponent in a position where he is controlled and can be moved.

Apply pressure with the foot against the bend in the opponent's leg above his calf. This lowers the opponent to the ground rather than throwing him to the ground and risking severe injury.

Striking Target Areas

Marines must avoid striking an opponent in the head, neck, or other bony parts with the baton because this is considered deadly force and can lead to serious bodily injury or death. Instead, the legs, arms, and buttocks are target areas that are considered nonlethal.

Legs. Primary targets are the thighs and lower legs. Avoid striking the knee and ankle joints because this can cause permanent damage.

Arms. Primary targets are the upper arms. Avoid striking the shoulder and elbow and wrist joints because this can cause permanent damage.

Buttocks. Primary targets are the buttocks. Avoid striking any other part of the torso, including the chest, rib cage, spine, tail bone, and groin because strikes to these areas can cause permanent damage or death.

One-Handed Striking Techniques

One-Handed Forward Strike. A forward strike follows either a horizontal line or a downward diagonal line using a forehand stroke. To execute the one-handed forward strike, Marines-

Stand facing the opponent with the baton carried in a one-handed cany.

Place the right hand palm up, swing the baton from right to left, and make contact with the opponent.

One-Handed Reverse Strike. A reverse strike follows either a horizontal line or a downward diagonal line using a backhand stroke. To execute the one-handed reverse strike, Marines—

Stand facing the opponent with the baton carried in a one-handed carry.

Bend the right arm and cross the arm to the left side of the body. The baton should be close to or over the left shoulder.

Place the right hand palm down, swing the baton from left to right, and make contact with the opponent.

Two-Handed Striking Techniques

Two-Handed Forward Strike. This strike is an effective follow-up to a middle block or left block. To execute the two-handed forward strike, Marines—

Pull back with the left hand while driving the right hand forward toward the opponent. The baton should be horizontal to the ground. Power is generated by stepping forward with the right foot and rotating the right hip and shoulder into the strike.

Contact the opponent with the end of the baton.

Two-Handed Reverse Strike. This strike is an effective follow-up to a middle block or left block. To execute the two-handed reverse strike, Marines—

Pull back with the right hand while driving the left hand forward toward the opponent. The baton should be horizontal to the ground. Power is generated by stepping forward slightly with the left foot and rotating the left hip and shoulder into the strike.

Contact the subject with the end of the baton.

Front Jab. This strike is effective for countering a frontal attack. It can also be executed as a quick poke to keep a subject away. To execute the front jab, Marines thrust both hands forward in a quick jab. The baton is held either horizontal to the ground or at a slight downward angle.

Rear Jab. This strike is effective for countering a bear hug from the rear. To execute the rear jab, Marines thrust both hands rearward in a quick jab. The baton is held either horizontal to the ground or at a slight downward angle.

APPENDIX A

PUGIL STICK TRAINING

A pugil stick is a training device used to simulate a rifle bayonet so that effective, but safe, training can be conducted to build proficiency of rifle bayonet techniques. Pugil stick training builds on the techniques used to throw punches. Pugil stick training is the only "full contact" training provided to Marines in the Close Combat Program. Pugil stick training teaches Marines to function when faced with stress and violence, and it prepares them to deliver a blow and take a blow. It also provides them with the physical and mental skills vital to success on the battlefield.

1. Pugil Stick Training

Design

A pugil stick consists of a stick wrapped in padding at both ends that can be gripped like a rifle. The pugil stick is approximately the same weight and length of an unloaded rifle with a bayonet attached.

Holding the Pugil Stick

The pugil stick is held in the same manner as the sen ice rifle. All movements come from the basic warrior stance. To hold the pugil stick correctly, Marines—

Use the right hand to grasp the lower end of the pugil stick overhanded.
Use the left hand to grasp the upper end of the pugil stick underhanded.
Use the right forearm to lock the lower end of the pugil stick against the hip.
Orient the blade end of the pugil stick toward the opponent.

Safety Equipment

The following safety equipment must be worn during any pugil stick bout.

Groin Protection. Groin protection protects the groin from an accidental blow. It should be pulled high around the waist to protect the groin area, with the concave portion against the body.

Flak Jacket. The flak jacket provides protection to the body. It is worn completely fastened.

Neck Roll. The neck roll prevents whiplash if Marines receive a blow to the head. The neck roll further supports the head and protects the neck from blows. The neck roll is worn above the flak jacket and below the helmet. The tied end of the neck roll faces the front.

Helmet. A regulation football helmet protects the face and head. The helmet must fit snugly and the chin straps must be adjusted and snapped.

Mouthpiece. The mouthpiece is worn on the upper teeth to protect the teeth.

Gloves. Marines wear gloves to protect the hands if pugil sticks do not have gloves built-in.

Pugil Stick Screening

Prior to pugil stick training, instructors must ask participating Marines the following questions. If a Marine answers "yes" to any question, he must see the corpsman or a competent medical authority who evaluates him and determines whether or not the he can participate in the training.

Are you on light duty?
Arc you restricted to running shoes by a corpsman or a doctor?
Have you fought in a pugil stick bout within the last 7 days?
Have you received a blow to the head within the last 7 days?
Have you had a concussion within the last 6 months?
Have you had dental surgery within the last 24 hours?
Do you have stitches or staples on your body?
Have you had a shoulder or head injury within the last 5 years?
Are you taking a prescription drug?
Do you have an ear infection or current sinus infection?
Have you had a broken bone within the last 6 months?

Safety Personnel

The following safety personnel are required to conduct pugil stick training:

One close combat instructor must officiate the bout.
One close combat instructor trainer, commissioned officer, or staff noncommissioned officer must be in the training area to serve as range safety officer.
One corpsman must be in the training area.

Training Area

To prevent injury, Marines train on areas with soft footing (i.e., sand or grass). Training mats are not recommended because feet can stick to the mats, prohibiting movement or causing joint injuries by twisting a knee. Bouts should not take place on a hard surface area; e.g., a flight deck or parking lot. A boxing ring may be used to conduct pugil stick bouts; ring dimensions can vary as long as there is ample room to execute the techniques in the training area.

Second Impact Syndrome

Second Impact Syndrome occurs when a second blow to the head produces a second concussion that occurs within 1 week following a previous concussion (before recovery from the first concussion). Second Impact Syndrome causes rapid brain swelling and can cause death. Therefore, there must be 7 days between pugil stick bouts to reduce the risk of severe injury resulting from Second Impact Syndrome. The 7 day separation between pugil stick bouts significantly reduces the possibility of injury, particularly in someone who may have suffered a brain injury or concussion but shows no symptoms.

Any Marine who experiences headaches or the following symptoms after training must be examined by appropriate medical personnel:

Blurred vision.
Ringing in the ears.
Dilation of the pupils.
Slurred speech.
Bleeding from ears or mouth.
Swelling in head or neck area.
Any unnatural discoloration in head or neck area.

The Marine should not be allowed to participate in pugil stick training or any other activity where a heavy blow might be sustained for a minimum of 7 days after the headache or other symptoms have subsided.

Safety Measures

The following safety measures must be followed:

Gear must be worn properly throughout training.
Contact lenses or glasses will not be worn.
False teeth will be removed from the mouth.
Nothing will be worn around the neck except the neck roll.

Competition among groups of Marines is authorized as long as it does not overshadow training objectives or compromise safety procedures.

Safety and proper techniques are paramount.

Safety is more important than competition.

2. General Rules and Regulations Governing Pugil Stick Bouts

Instructors and Support Personnel Requirements

At a minimum, one close combat instructor observes a pugil stick bout. For safety purposes, it is better to have two instructors judging a bout because each instructor can fully observe each of the fighters. The best position for observation is to the right of a fighter. This allows the instructor to see the fighter's facial expression and body movement. The instructor's position must not interfere with the fight.

In addition, one commissioned or staff noncommissioned officer (to serve as range safety officer) and one corpsman will be in the training area.

Prior to the Bout

Prior to the bout, Marines—

Are paired according to height, weight, and gender.

Wear the proper safety equipment.

Once Marines have properly donned the safety equipment, they wait for instructions from the close combat instructor. On the command of the close combat instructor, a pair of Marines will enter the ring. Once in the ring, the close combat instructor inspects each Marine for the proper safety equipment.

Note: There should he enough gear so that when two Marines are training, two other Marines can be donning safety equipment for the next bout.

During the Bout

The bout begins when the close combat instructor blows the whistle. All strikes are directed above the waist. Upon hearing a whistle blast, all fighting immediately ceases.

Stopping a Bout

There arc two reasons for stopping a bout: delivery of a scoring blow or an unsafe condition. The close combat instructor trainer, close combat instructor, or the range safety officer may stop the bout at any time an unsafe condition is observed.

Scoring Blow. A scoring blow is an offensive technique delivered to a vulnerable area of an opponent with sufficient force and precision to be considered as a disabling or killing blow. Scoring blows are not judged solely on the degree of force with which a blow is delivered, but on the accuracy and techniques employed. A scoring blow is defined as—

A straight thrust with the blade end of the weapon (red end of the pugil stick) to the opponent's face mask or throat.

A slash to the side of the opponent's helmet (below the ear) or neck with the red end of the pugil stick.

A heavy blow to the opponent's head with an authorized technique (i.e., buttstrokc, smash) using the butt (black end) of the pugil stick.

When a scoring blow is delivered, the close combat instructor blows the whistle to stop the bout.

Unsafe Condition. The bout will be stopped as soon as an unsafe condition exists. An unsafe condition exists when a Marine is unable to defend himself, loses his balance and falls down on one or two knees or falls down completely, shows instability (e.g., buckling at the knees), loses muscular tension in his neck and his head snaps back or to one side, or appears disoriented. An unsafe condition also exists if a Marine lets go of one end of the pugil stick, the equipment (e.g., helmet, neck roll) falls off, or a Marine fails to use the proper techniques.

If any of these conditions occur—

The close combat instructor trainer, close combat instructor, range safety officer, or anyone supervising training stops the bout and separates the two Marines.
The corpsman evaluates the possibly injured Marine to determine if the fight can continue. The corpsman—
- Ensures that he is alert and responsive, both verbally and physically.
- Talks to him to see if he is coherent.
- Makes certain he comprehends and replies to verbal questions.
- Makes certain his speech is not slurred.
- Checks his physical signs.
- Ensures his eyes are focused and not dazed or glazed.
- Ensures his legs are not wobbly or shaky.

The pugil stick is not used as a baseball bat. The use of unauthorized techniques will result in expulsion from the bout. Only techniques taught in bayonet training are authorized. These techniques include-

Straight thrust.
Buttstroke (horizontal and vertical).
Smash.
Slash.
Parry.
Blocks (low, high, left, right).

After the Bout

A third whistle blast is used prior to identifying the Marine who delivered the killing blow.

A fourth whistle blast is used prior to demonstrating the killing technique used by the Marine who won the bout.

3. Directions for Making Pugil Sticks

Close combat instructor trainers are taught to make pugil sticks to standard specifications at school. It is recommended that all pugil sticks be made under the supervision of a close combat instructor trainer. Until the Marine Corps acquires a standardized pugil stick that units can order through the supply system, instructors use the following information and instructions to make pugil sticks locally.

Supplies

The following supplies are needed to make one pugil stick:

One 45" by 2" circular, oak, wood dowel.
Four 9" wide by 27" long foam pad pieces to make two inner handguards of double thickness.
Four 11" wide by 19" long foam pad pieces to make two outer handguards of double thickness.
Canvas material to cover inner and outer handguards.
One foam pad cut 1/4" thick by 7" wide by 13" long to make the centerguard.
One 8" wide by 14"long piece of canvas material to make a pugil stick sock.
Canvas material to make two pugil stick end caps. Ideally, two different color fabrics should be used to simulate the bayonet and buttstock ends of a rifle. Common colors for end caps are black and red. Each end cap is a different color to help the officiating certified close combat instructor determine the scoring blow during the pugil stick bout.
Two 1/4" thick by 11" wide by 54" long foam pad pieces to make two ends.
Two foam pads cut 1/4" thick by 3" wide by 14 1/2" long to make two end plug inserts.
One can of 3M Photo MountQ spray adhesive (NSN 8040-01-120-4009)to adhere foam padding to the stick. (Spray adhesive is available from self-service.)
One roll of duct tape or riggers tape (NSN 7510-00-266-5016) to protect foam pad and

canvas from wear and to reinforce gluing action of spray adhesive. (Duct tape or riggers tape is available from self-service.)

One roll para-cord or 5-50 cord to tie both the inner and outer handguards together.

One roll of thread strong enough to sew canvas material together.

Sixteen grommets, one for each of the four corners of the inner and outer handguards.

One foam sleeping mat (NSN 8465-01-109-3369). Use unserviceable foam sleeping mats obtained from Defense Rcutilization and Marketing Office.

Tools

The follow ing tools are needed to make a pugil stick:

Tape measure, ruler, or yardstick.
Scissors.
Grommet machine.
Knife.
Sewing machine.

Directions

To make a pugil stick, Marines perform the following steps.

Preparing Materials

Step 1. Cut foam sleeping mat into the dimensions listed above. Cut mats to double thickness.

Step 2. Use spray adhesive to adhere the two 9" by 27" pieces of foam together. Use the spray adhesive to adhere the two 11" by 19" pieces of foam together. Let adhesive dry before using the double-thick foam for handguards.

Step 3. Cut canvas material to make two coverings for the inner handguards. Cut material to make two coverings for the outer handguards. Two inner handguards and two outer handguards are required for each pugil stick.

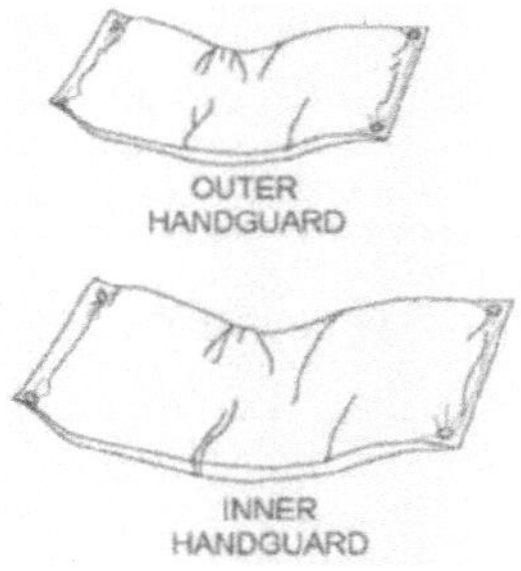

Step 4. Use a sewing machine to make a sleeve out of the 9" by 27" canvas. Sew down three sides. Insert foam pad inside the canvas sleeve and then sew down the fourth side. Use the grommet machine to create holes reinforced with metal grommets in each of the four corners. Once finished, the inner handguards arc complete.

Note: It is recommended that the Fabric Repair Shop be used to sew canvas sleeves and apply grommets.

Step 5. Repeat step 4 to make the other inner handguard and both outer handguards.

Constructing Ends, End Plugs, and End Caps

Step 1. Apply spray adhesive on approximately 8 inches of one end of the dowel. This adhesive serves to glue the foam pad to the wooden dowel.

Step 2. Apply spray adhesive on one side of the 11" by 54" foam pad. This causes the foam pad to stick to itself when it is rolled around the end of the wooden dowel. Ensure 3 inches of the rolled foam pad extends past the wooden dowel for stability.

Step 3. Apply spray adhesive on one side of 3" by 14 1/2" foam pad. Roll pad into a tight cylinder.

Step 4. Apply spray adhesive on the exposed end of the dowel and fill the hole in the end with the end plug. The end plug prevents the wooden dowel from protruding beyond the edge of the rolled foam padding. This is an important safety precaution to prevent injury.

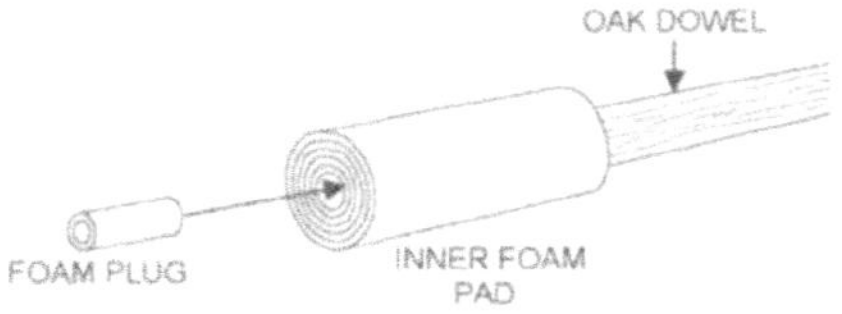

Step 5. Use duct tape or riggers tape to secure and reinforce the foam pad once the end plug is inserted in the end.

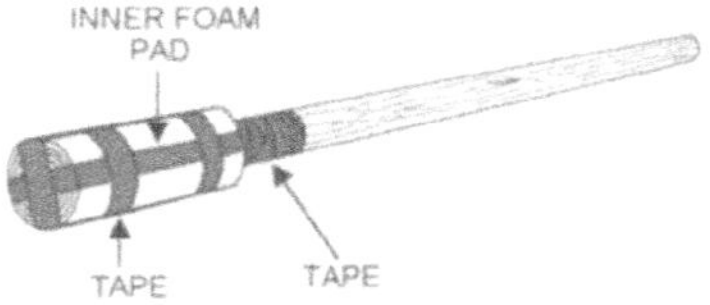

Step 6. Measure the dimensions of the ends (typically, the length of the cylinder should be 12 inches long to allow for taping). Cut and sew canvas material in the shape of a cylinder with one open end. Once complete, this becomes the end cap. The end cap is a colored sleeve that covers the rolled and taped foam pad at each end of the pugil stick.

Note: It is recommended that the Fabric Repair Shop be used to make the end cap.

Step 7. Slide one of the end caps over the rolled foam pad and tape it to the dowel.

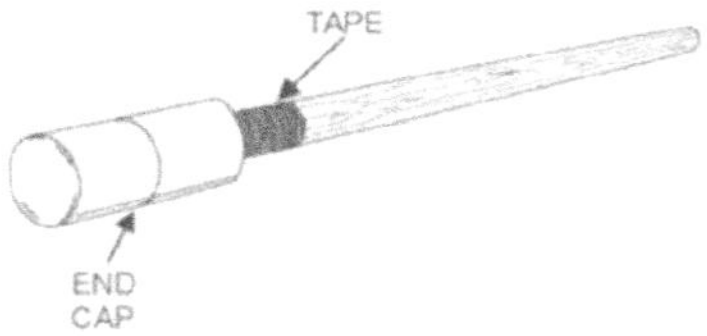

Building the Centerguard

Step 1. Spray one side of the 7" by 13" foam pad with spray adhesive and wrap it tightly around the center of the dowel.

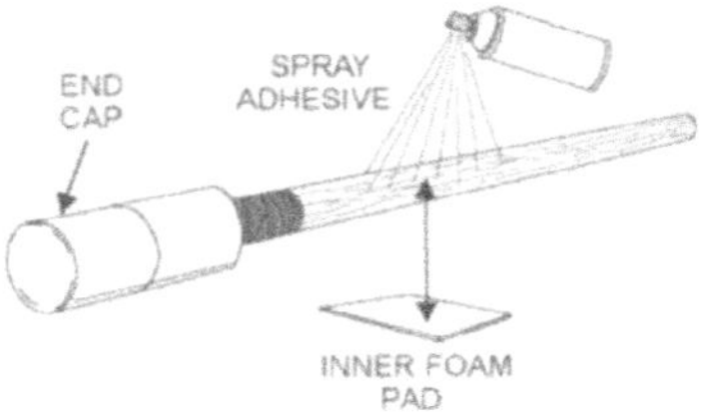

Step 2. Wrap duct tape or riggers tape around the foam pad to secure.

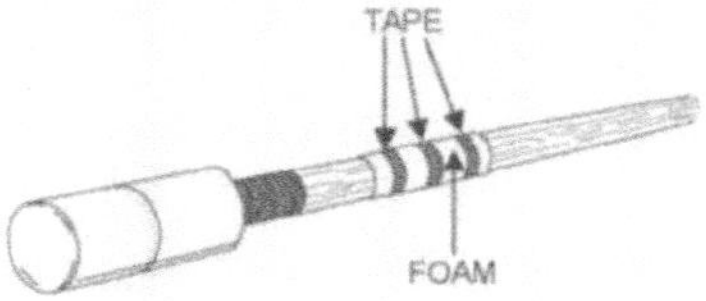

Step 3. Use a sewing machine to make a tubular pugil stick sock out of the 8" by 14" piece of canvas. Slide the pugil stick sock over the rolled foam pad and secure it with the duct tape or riggers tape. Duct tape or riggers tape protects the edges of the centerguard and reinforces the gluing effects of the spray adhesive.

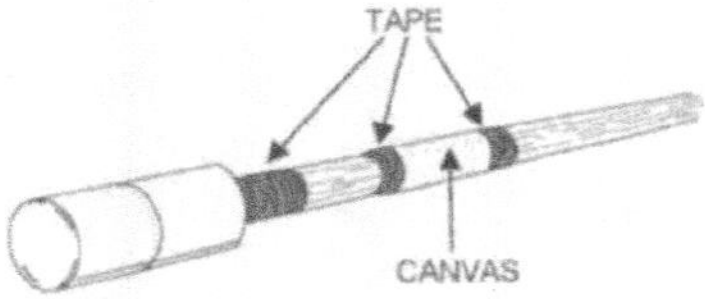

Note: It is recommended that the Fabric Repair Shop be used to make the pugil stick sock.

Finishing the Opposite End Cap

After one end cap is complete and the centerguard is finished, then repeat steps 1 through 7 in Constructing Ends, End Plugs, and End Caps for the other end of the pugil stick. Ensure that the opposite color end cap is used.

Fastening Handguards

Step 1. Place the inner handguards (longer piece) around the dowel. The inner handguard is placed on first because it protects the wrist. One inner handguard is needed for the left hand and one for the right hand.

Step 2. Fold the inner handguard around the pugil stick and run the para-cord or 5-50 cord through the metal grommets.

Step 3. Tighten down the corner ends of the inner handguards by pulling on the para-cord or 5-50 cord.

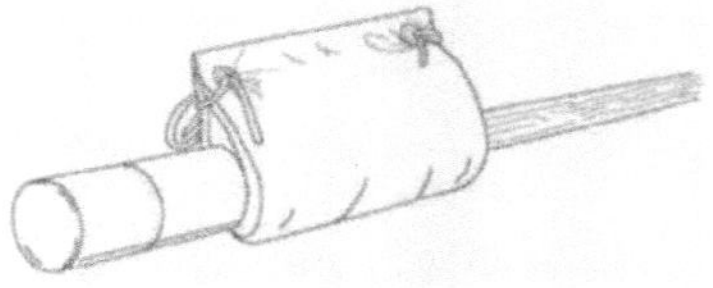

Step 4. Tie off the cord with a square knot above the metal grommets.

Step 5. Place the outer handguards (shorter piece) over the inner handguards (longer piece). Repeat steps 1-4 for the other handguard. The outer handguards provide additional protection to the knuckles and fingers and provides additional stability for the inner handguards.

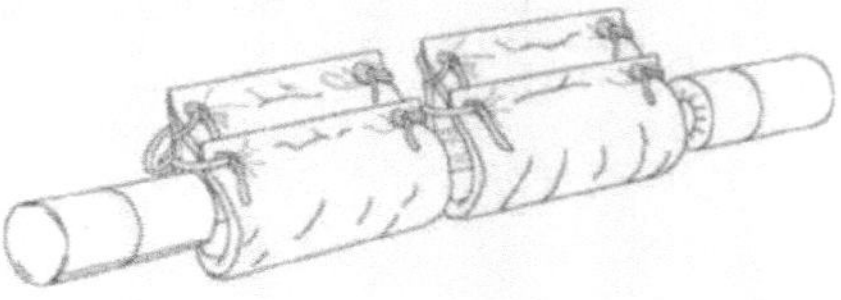

Serviceability Inspection

Prior to any training, all pugil sticks are inspected to ensure they are fit for use. If any of the following conditions exist, the pugil stick is unsafe and will not be used:

Stick protrudes through the end caps.
Frayed or worn edges exist on canvas surfaces.
The stick can be felt through the pads or the padding is too soft.
End caps or handguards are not securely attached.

APPENDIX B

SAFETY PRECAUTIONS DURING TRAINING

1. General Safety Precautions

When training close combat techniques, certain safety precautions must be adhered to to prevent injuries.

Most training should be conducted on a training area with soft footing such as a sandy or grassy area. If training mats are available, they should be used. A hard surface area is not appropriate for close combat training.

All techniques should be executed slowly at first. Marines can increase the speed of execution as they become more proficient. Marine-on-Marine training that requires contact (chokes, throws, ground fighting, and unarmed restraints and manipulation) should not be executed at full force or full speed.

If a technique is applied to the point that a Marine is uncomfortable, the Marine must "tap out." This indicates immediate release of the pressure being applied or to immediately stop the technique. The Marine "taps out" by firmly tapping his hand several times on any part of the opponent's body that will get his attention or by saying *stop.*

Second Impact Syndrome occurs when a second concussion develops within hours, days, or weeks following a prior concussion (and before recovery from the first concussion). Second Impact Syndrome causes rapid brain swelling and can cause death. Marines who experience headaches or other symptoms following training must be examined by appropriate medical personnel. These symptoms can include, but are not limited to, blurred vision, ringing in the ears, dialation of the pupils, bleeding from the ears or mouth, slurred speech, swelling in head or neck area, or any unnatural discoloration of head or neck. They should not be allowed to participate in pugil stick training or any other activity where a heavy blow might be sustained for a minimum of 7 days after the symptoms have subsided.

2. Safety Precautions for Individual Techniques

Falls

When training Marines to fall, they should progress from the ground, to a kneeling or squatting position, and then to a standing position. This ensures they are comfortable and understand the technique before progressing to executing falls from a higher profile. This instructional technique vastly reduces the risk of injury.

Strikes and Punches

When training Marines to strike and punch, they begin by executing the techniques "in the air." As they become more proficient, they execute strikes on equipment (when available) such as an air shield or a heavy bag. At no time should they be permitted to execute strikes on another student.

Chokes and Ground Fighting Chokes

When training Marines to execute chokes, they will not apply pressure to the opponent's throat during training because the trachea and windpipe can be crushed. During training, Marines should practice the proper procedures for blood chokes. They should never execute air chokes. At no time during training should a choke be applied at full force or full speed. No choke should be held for longer than 5 seconds.

Unarmed Restraints and Manipulation

When training Marines to execute unarmed restraints and manipulation techniques, they utilize slow and steady pressure. Never apply these techniques at full force or at full speed.

Pugil Stick Training

A blow to the head during training will bruise the brain. A second blow to the head can cause death. This is known as the Second Impact Syndrome. There must be 7 days between pugil stick training to prevent injury or death.

Desert Operations

Department of Defense

Pentagon
Publishing

* FM 90-3
FMFM 7-27

Headquarters
Department of the Army
The Marine Corps
Washington, DC
24 August 1993

FM 90-3
DESERT OPERATIONS

Table of Contents

PREFACE

INTRODUCTION

CHAPTER 1 - THE ENVIRONMENT AND ITS EFFECTS ON PERSONNEL AND EQUIPMENT

- SECTION I. THE ENVIRONMENT
- SECTION II. ENVIRONMENTAL EFFECTS ON PERSONNEL
- SECTION III. ENVIRONMENTAL EFFECTS ON EQUIPMENT

CHAPTER 2 - PREPARATIONS FOR DESERT OPERATIONS

SECTION I. FACTORS TO BE CONSIDERED FOR DESERT OPERATIONS

SECTION II. INDIVIDUAL TRAINING

SECTION III. UNIT TRAINING

CHAPTER 3 - OPERATIONS IN DESERT CONDITIONS

SECTION I. HOW THE DESERT ENVIRONMENT AFFECTS TACTICAL OPERATIONS

SECTION II. OFFENSIVE OPERATIONS

SECTION III. DEFENSIVE OPERATIONS

CHAPTER 4 - COMBAT SERVICE SUPPORT

SECTION I. BASE DEVELOPMENT PLAN

SECTION II. THEATER SUPPORT

SECTION III. CORPS SUPPORT COMMAND

SECTION IV. DIVISION SUPPORT COMMAND

SECTION V. COMBAT SERVICE SUPPORT ELEMENT

SECTION VI. SUPPORT OPERATIONS

SECTION VII. SECURITY OF SUPPLY ROUTES

SECTION VIII. SUPPLY

SECTION IX. MAINTENANCE

SECTION X. PERSONNEL SUPPORT

SECTION XI. HEALTH SERVICES

SECTION XII. NAVAL AND AIR FORCE ASSISTANCE

SECTION XIII. OTHER COMBAT SERVICE SUPPORT ISSUES

APPENDIX A - DESERT COUNTRIES OF THE WORLD

APPENDIX B - EMPLOYMENT OF AIRCRAFT IN DESERT OPERATIONS

APPENDIX C - TECHNIQUES FOR OPERATING EQUIPMENT IN THE DESERT

APPENDIX D - EFFECTS OF THE ENVIRONMENT ON NUCLEAR , BIOLOGICAL, AND CHEMICAL WEAPONS

APPENDIX E - DESERT CONCEALMENT AND CAMOUFLAGE

APPENDIX F - OPERATIONS IN MOUNTAINS

APPENDIX G - WATER USAGE IN DESERT OPERATIONS

APPENDIX H - FRATRICIDE REDUCTION

GLOSSARY

REFERENCES

AUTHORIZATION LETTER

DISTRIBUTION RESTRICTION: Approved for public release; distribution is unlimited.

* This publication supersedes FM 90-3, 19 August 1977.
U.S. Marine Corps: PCN 139 000670 00

INTRODUCTION

Arid regions make up about one-third of the earth's land surface, a higher percentage then that of any other type of climate. As we have seen in the recent past, some of these regions—because of diverse and conflicting culturs, strategic importance, and natural resources—have become centers of conflict.

Military leaders have long recognized the potential for US involvement in conflict In these regions. Exercises at the Array's National Training Center, Fort Irwin and the marine Corps' Marine Air Ground Combat Center, Twentynine Palms. California, have provided an opportunity for virtually all our ground forces to experience desert conditions, The success of Operation Desert Storm can be directly attributed to this realistic training.

Desert operations demand adaptation to the environment and to the limitations imposed by terrain and climate, Success depends on an appreciation of the effects of arid conditions on soldiers (both physically and psychologically), on equipment and facilities, and on eombot and support operations. Leaders and soldiers must continually evaluate the situation and be ready to react to changing conditions. Equipment and tactics must be modified and adapted to a dusty, rugged landscape where temperatures vary from extreme highs to freezing lows and where visibility can clange from 30 miles to 30 feet in a matter of minutes.

The key to success in desert operations is mobility. This was clearly evident in the ground operations of Desert Storm, The tactics employed to achieve victory over Iraq were wide, rapid flanking movements similar to those executed by Montgomery and Rommel during World War II, During Desert Storm, however, new technologies increased higher-echelon headquarter' ability to target, attack, and fight deep operations simultaneously. Modem weapon systems like the

M1 A1 Abrams tank, Bradley fighting vehicle, light armored vehicle, and assault amphibious vehicle, coupled with newly developed navigation and targeting devices, contributed immeasurably. Taclictl units were able to fight battles with minimal direction; leaders were able to exercise initiative based on a clear understanding of their commanders' intent. Current doctrine—Focused on improving mobility and implemented through, the planning, preparation, and execution processes, battle drills, and tactical SOPs, paved the way for the overwhelming triumph.

Arid regions create ***bold*** opportunities and resurants **for** soldiers and marines at all levels. The US military's performance in Desert Storm shows it understands these factors and has successfully addressed the effects of desert warfare TO troops, equipment, and operations. As they prepare for the future, leaders, soldiers, and marines must study past carnpaigns and use the lessons they learn **to** reduce casualties, use the environment to their advantage, and ensure victory on the desert battlefield.

CHAPTER 1

THE ENVIRONMENT AND ITS EFFECTS ON PERSONNEL AND EQUIPMENT

This chapter describes the desert environment and how it affects personnel and equipment.

Section I. The Environment

Successful desert operations require adaptation to the environment and to the limitations its terrain and climate impose. Equipment and tactics must be modified and adapted to a dusty and rugged landscape where temperatures vary from extreme highs down to freezing and where visibility may change from 30 miles to 30 feet in a matter of minutes. Deserts are arid, barren regions of the earth incapable of supporting normal life due to lack of water. See Figure 1-1 for arid regions of the world. Temperatures vary according to latitude and season, from over 136 degrees Fahrenheit in the deserts of Mexico and Libya to the bitter cold of winter in the Gobi (East Asia). In some deserts, day-to-night temperature fluctuation exceeds 70 degrees Fahrenheit. Some species of animal and plant life have adapted successfully to desert conditions where annual rainfall may vary from 0 to 10 inches.

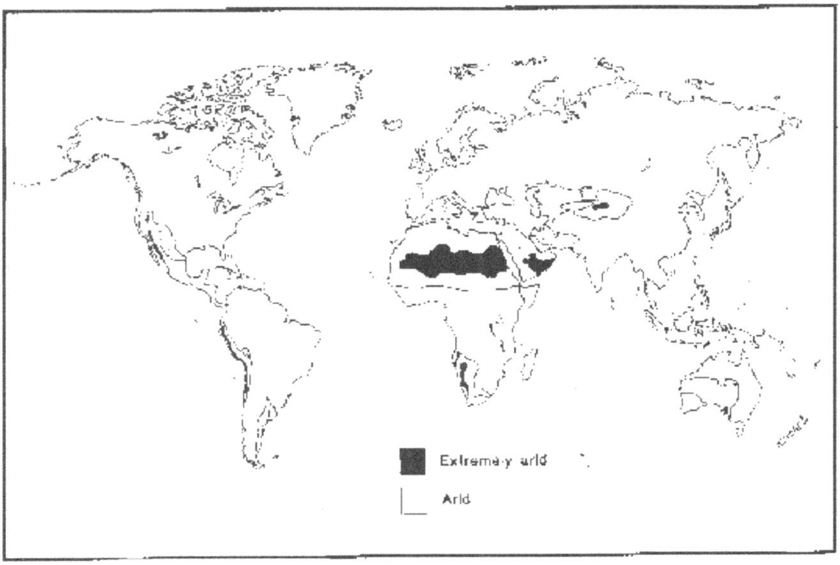

Figure 1-1. Desert of the world.

Desert terrain also varies considerably from place to place, the sole common denominator being lack of water with its consequent environmental effects, such as sparse, if any, vegetation. The basic land forms are similar to those in other parts of the world, but the topsoil has been eroded due to a combination of lack of water, heat, and wind to give deserts their characteristic barren appearance. The bedrock may be covered by a flat layer of sand, or gravel, or may have been exposed by erosion. Other common features are sand dunes, escarpments, wadis, and depressions. This environment can profoundly affect military operations. See Figure 1-2 for locations of major deserts of the world, and Appendix A for additional information on desert countries of the world.

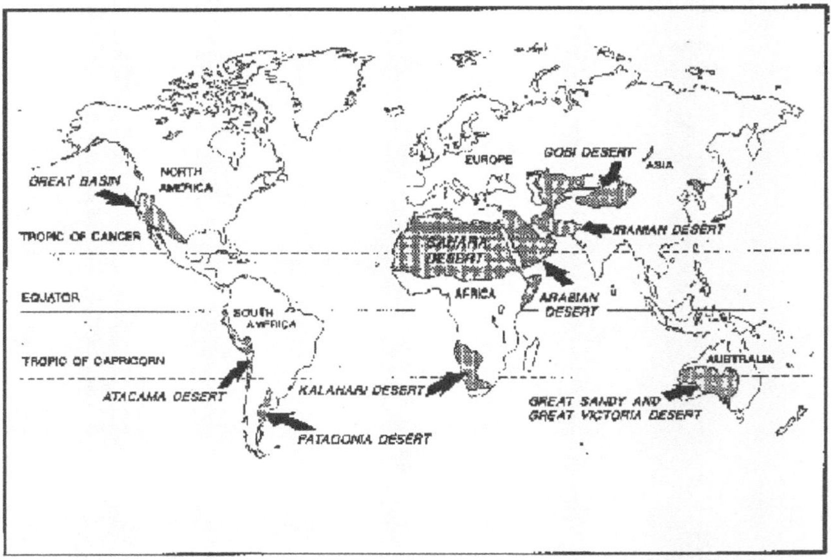

It is important to realize that deserts are affected by seasons. Those in the Southern Hemisphere have summer between 21 December and 21 March. This 6-month difference from the United States is important when considering equipping and training nonacclimatized soldiers/marines for desert operations south of the equator.

TERRAIN

Key terrain in the desert is largely dependent on the restrictions to movement that are present. If the desert floor will not support wheeled vehicle traffic, the few roads and desert tracks become key terrain. Crossroads are vital as they control military operations in a large area. Desert warfare is often a battle for control of the lines of communication (LOC). The side that can protect its own LOC while interdicting those of the enemy will prevail. Water sources are vital, especially if a force is incapable of long distance resupply of its water requirements. Defiles play an important role, where they exist. In the Western Desert of Libya, an escarpment that paralleled the coast was a barrier to movement except through a few passes. Control of these passes was vital. Similar escarpments are found in Saudi Arabia and Kuwait.

Types of Desert Terrain

There are three types of desert terrain: mountain, rocky plateau, and sandy or dune terrain. The following paragraphs discuss these types of terrain.

Mountain Deserts

Mountain deserts are characterized by scattered ranges or areas of barren hills or mountains, separated by dry, flat basins. See Figure 1-3 for an example of mountain desert terrain. High ground may rise gradually or abruptly from flat areas, to a height of several thousand feet above sea level. Most of the infrequent rainfall occurs on high ground and runs off in the form of flash floods, eroding deep gullies and ravines and depositing sand and gravel around the edges of the basins. Water evaporates rapidly, leaving the land as barren as before, although there may be short-lived vegetation. If sufficient water enters the basin to compensate for the rate of evaporation, shallow lakes may develop, such as the Great Salt Lake in Utah or the Dead Sea; most of these have a high salt content.

Figure 1-3, Example of mountain desert terrian,

Rocky Plateau Deserts

Rocky plateau deserts are extensive flat areas with quantities of solid or broken rock at or near the surface. See Figure 1-4 for an example of a rocky plateau desert. They may be wet or dry, steep-walled eroded valleys, known as wadis, gulches, or canyons. Narrow valleys can be extremely dangerous to men and materiel due to flash flooding after rains; although their flat bottoms may be superficially attractive as assembly areas. The National Training Center and the Golan Heights are examples of rocky plateau deserts.

figure1-4, Example of rocky plateau desert terrain.

Sandy or Dune Deserts

Sandy or dune deserts are extensive flat areas covered with sand or gravel, the product of ancient deposits or modern wind erosion. "Flat" is relative in this case, as some areas may contain sand dunes that are over 1,000 feet high and 10-15 miles long; trafficability on this type of terrain will depend on windward/leeward gradients of the dunes and the texture of the sand. See Figure 1-5 for an example of a sandy desert. Other areas, however, may be totally flat for distances of 3,000 meters and beyond. Plant life may vary from none to scrub, reaching over 6 feet high. Examples of this type of desert include the ergs of the Sahara, the Empty Quarter of the Arabian desert, areas of California and New Mexico, and the Kalahari in South Africa. See Figure 1-6 for an example of a dune desert.

figure 1 -3. Example of sandy desert terrain,

Figure 1-6. Example of dune desert terrain.

Trafficability

Roads and trails are rare in the open desert. Complex road systems beyond simple commercial links are not needed. Road systems have been used for centuries to connect centers of commerce, or important religious shrines such as Mecca and Medina in Saudi Arabia. These road systems are supplemented by routes joining oil or other mineral deposits to collection outlet points. Some surfaces, such as lava beds or salt marshes, preclude any form of routine vehicular movement, but generally ground movement is possible in all directions. Speed of movement varies depending on surface texture. Rudimentary trails are used by minor caravans and nomadic tribesmen, with wells or oases approximately every 20 to 40 miles; although there are some waterless stretches which extend over 100 miles. Trails vary in width from a few meters to over 800 meters.

Vehicle travel in mountainous desert country may be severely restricted. Available mutes can be easily blinked by the enemy or by climatic conditions. Hairpin turns are common on the edges of precipitous mountain gorges, and the higher passes may be blocked by snow in the winter.

Natural Factors

The following terrain features require special considerations regarding trafficability.

Wadis or dried water courses, vary from wide, but barely perceptible depressions of soft sand, dotted with bushes, to deep, steep-sided ravines. There frequently is a passable route through the bottom of a dried wadi. Wadis can provide cover from ground observation and camouflage from visual air reconnaissance. The threat of flash floods after heavy rains poses a significant danger to troops and equipment downstream. Flooding may occur in these areas even if it is not raining in the immediate area. See Figure 1-7 for an example of a wadi.

Figure 1-7. Example of a wadi

Salt marsh (sebkha) terrain is impassable to tracks and wheels when wet. When dry it has a brittle, crusty surface, negotiable by light wheel vehicles only. Salt marshes develop at points where the water in the subsoil of the desert rose to the surface. Because of the constant evaporation in the desert, the salts carried by the water are deposited, and results in a hard, brittle crust.

Salt marshes are normally impassable, the worst type being those with a dry crust of silt on top. Marsh mud used on desert sand will, however, produce an excellent temporary road. Many desert areas have salt marshes either in the center of a drainage basin or near the sea coast. Old trails or paths may cross the marsh, which are visible during the dry season but not in the wet season. In the wet season trails are indicated by standing water due to the crust being too hard or too thick for it to penetrate. However, such routes should not be tried by load-carrying vehicles without prior reconnaissance and marking. Vehicles may become mired so severely as to render equipment and units combat ineffective. Heavier track-laying vehicles, like tanks, are especially susceptible to these areas, therefore reconnaissance is critical.

Man-made Factors

The ruins of earlier civilizations, scattered across the deserts of the world, often are sited along important avenues of approach and frequently dominate the only available passes in difficult terrain. Control of these positions maybe imperative for any force intending to dominate the immediate area. Currently occupied dwellings have little impact on trafficability except that they are normally located near roads and trails. Apart from nomadic tribesmen who live in tents (see Figure 1-8 for an example of desert nomads), the population lives in thick-walled structures with small windows, usually built of masonry or a mud and straw (adobe) mixture. Figure 1-9 shows common man-made desert structures.

Figure 1-8. Example of desert nomads

Figure 1-9. Common man_made desert.

Because of exploration for and production of oil and other resources, wells, pipelines, refineries, quarries, and crushing plants may be of strategic importance in the desert. Pipelines are often raised 1 meter off the ground-where this is the case, pipelines will inhibit movement. Subsurface pipelines can also be an obstacle. In Southwest Asia, the subsurface pipelines were indicated on maps. Often they were buried at such a shallow depth that they could be damaged by heavy vehicles traversing them. Furthermore, if a pipeline is ruptured, not only is the spill of oil a consideration, but the fumes maybe hazardous as well.

Agriculture in desert areas has little effect on trafficability except that canals limit surface mobility. Destruction of an irrigation system, which may be a result of military operations, could have a devastating effect on the local population and should be an important consideration in operational estimates. Figure 1-10 shows an irrigation ditch.

Figure 1-10. Irrigation ditch

TEMPERATURE

The highest known ambient temperature recorded in a desert was 136 degrees Fahrenheit (58 degrees Celsius). Lower temperatures than this produced internal tank temperatures approaching 160 degrees Fahrenheit (71 degrees Celsius) in the Sahara Desert during the Second World War. Winter temperatures in Siberian deserts and in the Gobi reach minus 50 degrees Fahrenheit (minus 45 degrees Celsius). Low temperatures are aggravated by very strong winds producing high windchill factors. The cloudless sky of the desert permits the earth to heat during sunlit hours, yet cool to near freezing at night. In the inland Sinai, for example, day-to-night temperature fluctuations are as much as 72 degrees Fahrenheit.

WINDS

Desert winds can achieve velocities of near hurricane force; dust and sand suspended within them make life intolerable, maintenance very difficult, and restrict visibility to a few meters. The Sahara "Khamseen", for example, lasts for days at a time; although it normally only occurs in the spring and summer. The deserts of Iran are equally well known for the "wind of 120 days," with sand blowing almost constantly from the north at wind velocities of up to 75 miles per hour.

Although there is no danger of a man being buried alive by a sandstorm, individuals can become separated from their units. In all deserts, rapid temperature changes invariably follow strong winds. Even without wind, the telltale clouds raised by wheels, tracks, and marching troops give away movement. Wind aggravates the problem. As the day gets warmer the wind increases and the dust signatures of vehicles may drift downwind for several hundred meters.

In the evening the wind normally settles down. In many deserts a prevailing wind blows steadily from one cardinal direction for most of the year, and eventually switches to another direction for the remaining months. The equinoctial gales raise huge sandstorms that rise to several thousand feet and may last for several days. Gales and sandstorms in the winter months can be bitterly cold. See Figure 1-11 for an example of wind erosion.

Figure 1-11. Example of wind crosion

Sandstorms are likely to form suddenly and stop just as suddenly. In a severe sandstorm, sand permeates everything making movement nearly impossible, not only because of limited visibility, but also because blowing sand damages moving parts of machinery.

WATER

The lack of water is the most important single characteristic of the desert. The population, if any, varies directly with local water supply. A Sahara oasis may, for its size, be one of the most densely occupied places on earth (see Figure 1-12 for a typical oasis).

Figure 1-12. Typical oasis

Desert rainfall varies from one day in the year to intermittent showers throughout the winter. Severe thunderstorms bring heavy rain, and usually far too much rain falls far too quickly to organize collection on a systematic basis. The water soon soaks into the ground and may result in flash floods. In some cases the rain binds the sand much like a beach after the tide ebbs allowing easy maneuver however, it also turns loam into an impassable quagmire obstacle. Rainstorms tend to be localized, affecting only a few square kilometers at a time. Whenever possible, as storms approach, vehicles should move to rocky areas or high ground to avoid flash floods and becoming mired.

Permanent rivers such as the Nile, the Colorado, or the Kuiseb in the Namib Desert of Southwest Africa are fed by heavy precipitation outside the desert so the river survives despite a high evaporation rate.

Subsurface water may be so far below the surface, or so limited, that wells are normally inadequate to support any great number of people. Because potable water is absolutely vital, a large natural supply may be both tactically and strategically important. Destruction of a water supply system may become a political rather than military decision, because of its lasting effects on the resident civilian population.

Finding Water

When there is no surface water, tap into the earth's water table for ground water. Access to this table and its supply of generally pure water depends on the contour of the land and the type of soil. See Figure 1-13 for water tables.

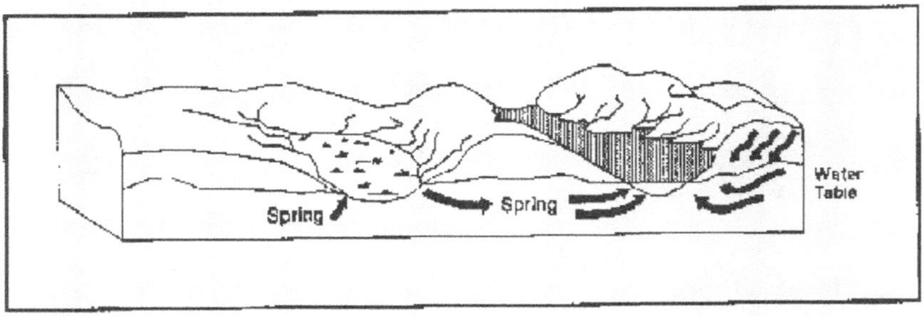

Figure 1-13. Water tables.

From Rocky Soil

Look for springs and seepages. Limestone has more and larger springs than any other type rock. Because limestone is easily dissolved, caverns are readily etched in it by ground water. Look in these caverns for springs. Lava rock is a good source of seeping ground water because it is porous. Look for springs along the walls of valleys that cross the lava flow. Look for seepage where a dry canyon cuts through a layer of porous sandstone.

Watch for water indicators in desert environments. Some signs to look for are the direction in which certain birds fly, the location of plants, and the convergence of game trails. Asian sand grouse, crested larks, and zebra birds visit water holes at least once a day. Parrots and pigeons must live within reach of water. Cattails, greasewoods, willows, elderberry, rushes, and salt grass grow only where ground water is near the surface. Look for these signs and dig. If you do not have a bayonet or entrenching tool, dig with a flat rock or sharp stick.

Desert natives often know of lingering surface pools in low places. They cover their surface pools, so look under brush heaps or in sheltered nooks, especially in semiarid and brush country.

Places that are visibly damp, where animals have scratched, or where flies hover, indicate recent surface water. Dig in such places for water. Collect dew on clear nights by sponging it up with a handkerchief. During a heavy dew you should be able to collect about a pint an hour.

Dig in dry stream beds because water may be found under the gravel. When in snow fields, put in a water container and place it in the sun out of the wind.

From Plants

If unsuccessful in your search for ground or runoff water, or if you do not have time to purify the questionable water, a water-yielding plant may be the best source. Clear sap from many plants is easily obtained. This sap is pure and is mostly water.

Plant tissues . Many plants with fleshy leaves or stems store drinkable water. Try them wherever you find them. The barrel cactus of the southwestern United States is a possible source of water (see Figure 1-14). Use it only as a last resort and only if you have the energy to cut through the tough, spine-studded outer rind. Cut off the top of the cactus and smash the pulp within the plant. Catch the liquid in a container. Chunks may be carried as an emergency water source. A barrel cactus 3-1/2 feet high will yield about a quart of milky juice and is an exception to the rule that milky or colored sap-bearing plants should not be eaten.

figure 1-14. Barrel cactus of possible source of water.

Roots of desert plants. Desert plants often have their roots near the surface. The Australian water tree, desert oak, and bloodwood are some examples. Pry these roots out of the ground, cut them into 24-36 inch lengths, remove the bark, and suck the water.

Vines . Not all vines yield palatable water, but try any vine found. Use the following method for tapping a vine--it will work on any species:

Step 1. Cut a deep notch in the vine as high up as you can reach.

Step 2. Cut the vine off close to the ground and let the water drip into your mouth or into a container.

Step 3. When the water ceases to drip, cut another section off the top. Repeat this until the supply of fluid is exhausted.

Palms. Burl, coconut, sugar and nipa palms contain a drinkable sugary fluid. To start the flow in coconut palm, bend the flower stalk downward and cut off the top. If a thin slice is cut off the stalk every 12 hours, you can renew the flow and collect up to a quart a day.

Coconut. Select green coconuts. They can be opened easily with a knife and they have more milk than ripe coconuts. The juice of a ripe coconut is extremely laxative; therefore, do not drink more than three or four cups a day.

The milk of a coconut can be obtained by piercing two eyes of the coconut with a sharp object such as a stick or a nail. To break off the outer fibrous covering of the coconut without a knife, slam the coconut forcefully on the point of a rock or protruding stump.

Survival Water Still

You can build a cheap and simple survival still that will produce drinking water in a dry desert. Basic materials for setting up this still are-

- 6-foot square sheet of clean plastic.
- A 2- to 4-quart capacity container.
- A 5-foot piece of flexible plastic tubing.

Pick an unshaded spot for the still, and dig a hole. If no shovel is available, use a stick or even your hands. The hole should be about 3 feet across for a few inches down, then slope the hole toward the bottom as shown in Figure 1-15 which depicts a cross section of a survival still. The hole should be deep enough so the point of the plastic cone will be about 18 inches below ground and will still clear the top of the container. Once the hole is properly dug, tape one end of the plastic drinking tube inside the container and center the container in the bottom of the hole. Leave the top end of the drinking tube free, lay the plastic sheet over the hole, and pile enough dirt around the edge of the plastic to hold it securely. Use a fist-size rock to weight down the center of the plastic; adjust the plastic as necessary to bring it within a couple of inches of the top of the container. Heat from the sun vaporizes the ground water. This vapor condenses under the plastic, trickles down, and drops into the container.

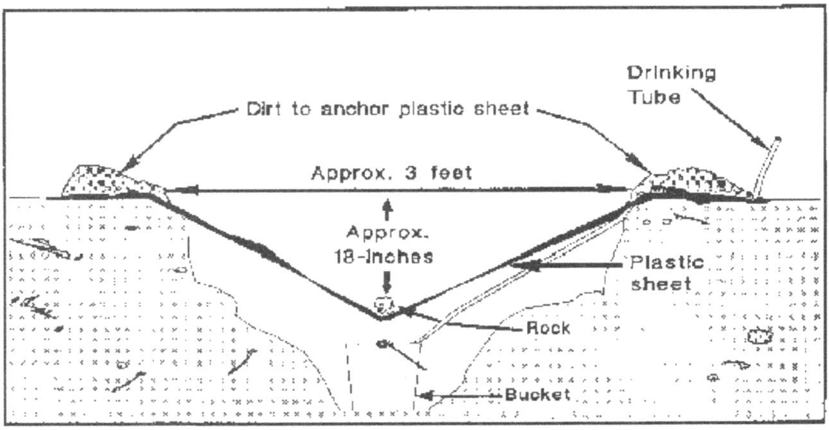

Figure 1-15. Cross-section of a survival still.

VEGETATION

The indigenous vegetation and wildlife of a desert have physiologically adapted to the conditions of the desert environment. For example, the cacti of the American desert store moisture in enlarged stems. Some plants have drought-resistant seeds that may lie dormant for years, followed by a brief, but colorful display of growth after a rainstorm. The available vegetation is usually inadequate to provide much shade, shelter, or concealment, especially from the air. Some plants, like the desert gourd, have vines which grow to 4.5 meters (15 feet). Others have wide lateral roots just below the surface to take advantage of rain and dew, while still others grow deep roots to tap subsurface water. Presence of palm trees usually indicates water within a meter of the surface, salt grass within 2 meters, Cottonwood and willows up to 4 meters. In addition to indicating the presence of water, some plants are edible.

WILDLIFE

Invertebrates such as ground-dwelling spiders, scorpions, and centipedes, together with insects of almost every type, are in the desert. Drawn to man as a source of moisture or food, lice, mites, and flies can be extremely unpleasant and carry diseases such as scrub typhus and dysentery. The stings of scorpions and the bites of centipedes and spiders are extremely painful, though seldom fatal. Some species of scorpion, as well as black widow and recluse spiders, can cause death. The following paragraphs describe some of the wildlife that are encountered in desert areas and the hazards they may pose to man.

Scorpions

Scorpions are prevalent in desert regions, particularly active at night. Scorpions are They prefer damp locations and are easily recognizable by their crab-like appearance, and by their long tail which ends in a sharp stinger. Adult scorpions vary from less than an inch to almost 8 inches in length. Colors range from nearly black to straw to striped. Scorpions hide in clothing, boots, or bedding, so troops should routinely shake these items before using. Although scorpion stings are rarely fatal, they can be painful.

Flies

Flies are abundant throughout desert environments. Filth-borne disease is a major health problem posed by flies. Dirt or insects in the desert can cause infection in minor cuts and scratches.

Fleas

Avoid all dogs and rats which are the major carriers of fleas. Fleas are the primary carriers of plague and murine typhus.

Reptiles

Reptiles are perhaps the most characteristic group of desert animals. Lizards and snakes occur in quantity, and crocodiles are common in some desert rivers. Lizards are normally harmless and can be ignored; although exceptions occur in North America and Saudi Arabia.

Snakes, ranging from the totally harmless to the lethal, abound in the desert. A bite from a poisonous snake under two feet long can easily become infected. Snakes seek shade (cool areas) under bushes, rocks, trees, and shrubs. These areas should be checked before sitting or resting. Troops should always check clothing and boots before putting them on. Vehicle operators should look for snakes when initially conducting before-operations maintenance. Look for snakes in and around suspension components and engine compartments as snakes may seek the warm areas on recently parked vehicles to avoid the cool night temperatures.

Sand vipers have two long and distinctive fangs that may be covered with a curtain of flesh or folded back into the mouth. Sand vipers usually are aggressive and dangerous in spite of their size. A sand viper usually buries itself in the sand and may strike at a passing man; its presence is alerted by a characteristic coiling pattern left on the sand.

The Egyptian cobra can be identified by its characteristic cobra combative posture. In this posture, the upper portion of the body is raised vertically and the head tilted sharply forward. The neck is usually flattened to form a hood. The Egyptian cobra is often found around rocky places and ruins and is fairly common. The distance the cobra can strike in a forward direction is equal to the distance the head is raised above the ground. Poking around in holes and rock piles is particularly dangerous because of the likelihood of encountering a cobra. See Figure 1-16 for an example of a viper and cobra.

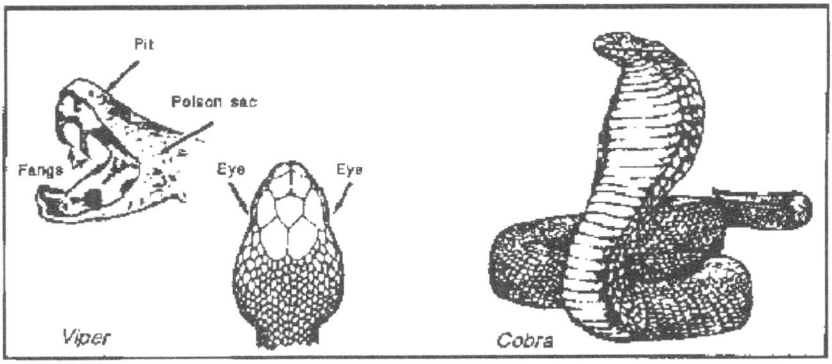

Figure 1-6. Sand viper and cobra

Mammals

The camel is the best known desert mammal. The urine of the camel is very concentrated to reduce water loss, allowing it to lose 30 percent of its body weight without undue distress. A proportionate loss would be fatal to man. The camel regains this weight by drinking up to 27 gallons (120 liters) of water at a time. It cannot, however, live indefinitely without water and will die of dehydration as readily as man in equivalent circumstances. Other mammals, such as gazelles, obtain most of their required water supply from the vegetation they eat and live in areas where there is no open water. Smaller animals, including rodents, conserve their moisture by burrowing underground away from the direct heat of the sun, only emerging for foraging at night. All these living things have adapted to the environment over a period of thousands of years; however, man has not made this adaptation and must carry his food and water with him and must also adapt to this severe environment.

Dogs are often found near mess facilities and tend to be in packs of 8 or 10. Dogs are carriers of rabies and should be avoided. Commanders must decide how to deal with packs of dogs; extermination and avoidance are two options. Dogs also carry fleas which may be transferred upon bodily contact. Rabies is present in most desert mammal populations. Do not take any chances of contracting fleas or rabies from any animal by adopting pets.

Rats are carriers of various parasites and gastrointestinal diseases due to their presence in unsanitary locations.

Section II. Environmental Effects on Personnel

There is no reason to fear the desert environment, and it should not adversely affect the morale of a soldier/marine who is prepared for it. Lack of natural concealment has been known to induce temporary agoraphobia (fear of open spaces) in some troops new to desert conditions, but this fear normally disappears with acclimatization. Remember that there is nothing unique about either living or fighting in deserts; native tribesmen have lived in the Sahara for thousands of years. The British maintained a field army and won a campaign in the Western Desert in World War II at the far end of a 12,000-mile sea line of communication with equipment considerably inferior to that in service now. The desert is neutral, and affects both sides equally; the side whose personnel are best prepared for desert operations has a distinct advantage.

The desert is fatiguing, both physically and mentally. A high standard of discipline is essential, as a single individual's lapse may cause serious damage to his unit or to himself. Commanders must exercise a high level of leadership and train their subordinate leaders to assume greater responsibilities required by the wide dispersion of units common in desert warfare. Soldiers/marines with good leaders are more apt to accept heavy physical exertion and uncomfortable conditions. Every soldier/marine must clearly understand why he is fighting in such harsh conditions and should be kept informed of the operational situation. Ultimately, however, the maintenance of discipline will depend on individual training.

Commanders must pay special attention to the welfare of troops operating in the desert, as troops are unable to find any "comforts" except those provided by the command. Welfare is an essential factor in the maintenance of morale in a harsh environment, especially to the inexperienced. There is more to welfare than the provision of mail and clean clothing. Troops must be kept healthy and physically fit; they must have adequate, palatable, regular food, and be allowed periods of rest and sleep. These things will not always be possible and discomfort is inevitable, but if troops know that their commanders are doing everything they can to make life tolerable, they will more readily accept the extremes brought on by the environment.

HEAT

The extreme heat of the desert can cause heat exhaustion and heatstroke and puts troops at risk of degraded performance. For optimum mental and physical performance, body temperatures must be maintained within narrow limits. Thus, it is important that the body lose the heat it gains during work. The amount of heat accumulation in the human body depends upon the amount of physical activity, level of hydration, and the state of personal heat acclimatization. Unit leaders must monitor their troops carefully for signs of heat distress and adjust schedules, work rates, rest, and water consumption according to conditions.

Normally, several physical and physiological mechanisms (e.g., convection and evaporation) assure transfer of excess body heat to the air. But when air temperature is above skin temperature (around 92 degrees Fahrenheit) the evaporation of sweat is the only operative mechanism. Following the loss of sweat, water must be consumed to replace the body's lost fluids. If the body fluid lost through sweating is not replaced, dehydration will follow. This will hamper heat dissipation and can lead to heat illness. When humidity is high, evaporation of sweat is inhibited and there is a greater risk of dehydration or heat stress. Consider the following to help prevent dehydration:

- Heat, wind, and dry air combine to produce a higher individual water requirement, primarily through loss of body water as sweat. Sweat rates can be high even when the skin looks and feels dry.
- Dehydration nullifies the benefits of heat acclimatization and physical fitness, it increases the susceptibility to heat injury, reduces the capacity to work, and decreases appetite and alertness. A lack of alertness can indicate early stages of dehydration.

- Thirst is not an adequate indicator of dehydration. The soldier/marine will not sense when he is dehydrated and will fail to replace body water losses, even when drinking water is available. The universal experience in the desert is that troops exhibit "voluntary dehydration" that is, they maintain their hydration status at about 2 percent of body weight (1.5 quarts) below their ideal hydration status without any sense of thirst.

Chronic dehydration increases the incidence of several medical problems: constipation (already an issue in any field situation), piles (hemorrhoids), kidney stones, and urinary infections. The likelihood of these problems occurring can be reduced by enforcing mandatory drinking schedules.

Resting on hot sand will increase heat stress-the more a body surface is in contact with the sand, the greater the heat stress. Ground or sand in full sun is hot, usually 30-45 degrees hotter than the air, and may reach 150 degrees Fahrenheit when the air temperature is 120 degrees Fahrenheit. Cooler sand is just inches below the surface; a shaded trench will provide a cool resting spot.
At the first evidence of heat illness, have the troops stop work, get into shade, and rehydrate. Early intervention is important. Soldiers/ marines who are not taken care of can become more serious casualties.

WARNING

One heat casually Is usually followed by Others and Is a warning that the entire unit may be at risk, This is the "Week Link Rule." The slalus Of the whole unit Is assessed at this point.

ACCLIMATIZATION

Acclimatization to heat is necessary to permit the body to reach and maintain efficiency in its cooling process. A period of approximately 2 weeks should be allowed for acclimatization, with progressive increases in heat exposure and physical exertion. Significant acclimatization can be attained in 4-5 days, but full acclimatization takes 7-14 days, with 2-3 hours per day of exercise in the heat. Gradually increase physical activity until full acclimatization is achieved.

Acclimatization does not reduce, and may increase, water requirements. Although this strengthens heat resistance, there is no such thing as total protection against the debilitating effects of heat. Situations may arise where it is not possible for men to become fully acclimatized before being required to do heavy labor. When this happens heavy activity should be limited to cooler hours and troops should be allowed to rest frequently. Check the weather daily. Day-to-day and region-to-region variations in temperatures, wind, and humidity can be substantial.

CLIMATIC STRESS

Climatic stress on the human body in hot deserts can be caused by any combination of air temperature, humidity, air movement, and radiant heat. The body is also adversely affected by such factors as lack of acclimatization, being overweight, dehydration, alcohol consumption, lack of sleep, old age, and poor health.

The body maintains its optimum temperature of 98.6 degrees Fahrenheit by conduction/convection, radiation, and evaporation (sweat). The most important of these in the daytime desert is evaporation, as air temperature alone is probably already above skin temperature. If, however, relative humidity is high, air will not easily evaporate sweat and the cooling effect is reduced. The following paragraphs describe the effects of radiant light, wind, and sand on personnel in desert areas.

Radiant Light

Radiant light comes from all directions. The sun's rays, either direct or reflected off the ground, affect the skin and can also produce eyestrain and temporarily impaired vision. Not only does glare damage the eyes but it is very tiring; therefore, dark glasses or goggles should be worn.

Overexposure to the sun can cause sunburn. Persons with fair skin, freckled skin, ruddy complexions, or red hair are more susceptible to sunburn than others, but all personnel are susceptible to some degree. Personnel with darker complexions can also sunburn. This is difficult to monitor due to skin pigmentation, so leaders must be ever vigilant to watch for possible sunburn victims. Sunburn is characterized by painful reddened skin, and can result in blistering and lead to other forms of heat illness.

Soldier/marines should acquire a suntan in gradual stages (preferably in the early morning or late afternoon) to gain some protection against sunburn. They should not be permitted to expose bare skin to the sun for longer than five minutes on the first day, increasing exposure gradually at the rate of five minutes per day. They should be fully clothed in loose garments in all operational situations. This will also reduce sweat loss. It is important to remember that—

- The sun is as dangerous on cloudy days as it is on sunny days.
- Sunburn ointment is not designed to give complete protection against excessive exposure.
- Sunbathing or dozing in the desert sun can be fatal.

Wind

The wind can be as physically demanding as the heat, burning the face, arms, and any exposed skin with blown sand. Sand gets into eyes, nose, mouth, throat, lungs, ears, and hair, and reaches every part of the body. Even speaking and listening can be difficult. Continual exposure to blown sand is exhausting and demoralizing. Technical work spaces that are protected from dust and sand are likely to be very hot. Work/rest cycles and enforced water consumption will be required.

The combination of wind and dust or sand can cause extreme irritation to mucous membranes, chap the lips and other exposed skin surfaces, and can cause nosebleed. Cracked, chapped lips make eating difficult and cause communication problems. Irritative conjunctivitis, caused when fine particles enter the eyes, is a frequent complaint of vehicle crews, even those wearing goggles. Lip balm and skin and eye ointments must be used by all personnel. Constant wind noise is tiresome and increases soldier/marine fatigue, thus affecting alertness.

When visibility is reduced by sandstorms to the extent that military operations are impossible, soldiers/marines should not be allowed to leave their group for any purpose unless secured by lines for recovery.

The following are special considerations when performing operations in dust or sand:

- Contact lenses are very difficult to maintain in the dry dusty environment of the desert and should not be worn except by military personnel operating in air conditioned environments, under command guidance.
- Mucous membranes can be protected by breathing through a wet face cloth, snuffing small amounts of water into nostrils (native water is not safe for this purpose) or coating the nostrils with

a small amount of petroleum jelly. Lips should be protected by lip balm.

- Moving vehicles create their own sandstorms and troops traveling in open vehicles should be protected.
- Scarves and bandannas can be used to protect the head and face.
- The face should be washed as often as possible. The eyelids should be cleaned daily.

BASIC HEAT INJURY PREVENTION

The temperature of the body is regulated within very narrow limits. Too little salt causes heat cramps; too little salt and insufficient water causes heat exhaustion. Heat exhaustion will cause a general collapse of the body's cooling mechanism. This condition is heatstroke, and is potentially fatal. To avoid these illnesses, troops should maintain their physical fitness by eating adequately, drinking sufficient water, and consuming adequate salt. If soldiers/marines expend more calories than they take in, they will be more prone to heat illnesses. Since troops may lose their desire for food in hot climates, they must be encouraged to eat, with the heavier meal of the day scheduled during the cooler hours.

It is necessary to recognize heat stress symptoms quickly. When suffering from heatstroke, the most dangerous condition, there is a tendency for a soldier/marine to creep away from his comrades and attempt to hide in a shady and secluded spot; if not found and treated, he will die. When shade is required during the day, it can best be provided by tarpaulins or camouflage nets, preferably doubled to allow air circulation between layers and dampened with any surplus water.

WARNING

Drink water even if not thirsty. Thirst Is not an adequate warning of dehydration.

Approximately 75 percent of the human body is fluid. All chemical activities in the body occur in a water solution, which assists in the removal of toxic body wastes and plays a vital part in the maintenance of an even body temperature. A loss of 2 quarts of body fluid (2.5 percent of body weight) decreases efficiency by 25 percent and a loss of fluid equal to 15 percent of body weight is usually fatal. The following are some considerations when operating in a desert environment:

- Consider water a tactical weapon. Reduce heat injury by forcing water consumption. Soldiers/marines in armored vehicles, MOPP gear, and in body armor need to increase their water intake.
- When possible, drink Cool (50-55 degrees Fahrenheit) water.
- Drink one quart of water in the morning, at each meal, and before strenuous work. In hot climates drink at least one quart of water each hour. At higher temperatures hourly water requirements increase to over two quarts.
- Take frequent drinks since they are more often effective than drinking the same amount all at once. Larger soldiers/marines need more water.
- Replace salt loss through eating meals.
- When possible, work loads and/or duration of physical activity should be less during the first days of exposure to heat, and then should gradually be increased to follow acclimatization.

- Modify activities when conditions that increase the risk of heat injury (fatigue/loss of sleep, previous heat exhaustion, taking medication) are present.
- Take frequent rest periods in the shade, if possible. Lower the work rate and work loads as the heat condition increases.
- Perform heavy work in the cooler hours of the day such as early morning or late evening, if possible.

A description of the symptoms and treatment for heat illnesses follows:

- Heat cramps.

  - Symptoms: Muscle cramps of arms, legs, and/or stomach. Heavy sweating (wet skin) and extreme thirst.

  - First aid: Move soldier/marine to a shady area and loosen clothing. Slowly give large amounts of cool water. Watch the soldier/marine and continue to give him water, if he accepts it. Get medical help if cramps continue.

- Heat exhaustion.

  - Symptoms: Heavy sweating with pale, moist, cool skin; headache, weakness, dizziness, and/or loss of appetite; heat cramps, nausea (with or without vomiting), rapid breathing, confusion, and tingling of the hands and/or feet.

  - First aid: Move the soldier/marine to a cool, shady area and loosen/remove clothing. Pour water on the soldier/marine and fan him to increase the cooling effect. Have the soldier/marine slowly drink at least one full canteen of water. Elevate the soldier's/marine's legs. Get medical help if symptoms continue; watch the soldier/marine until the symptoms are gone or medical aid arrives.

- Heatstroke.

  - Symptoms: Sweating stops (red, flushed, hot dry skin).

  - First aid: Evacuate to a medical facility immediately. Move the soldier/marine to a cool, shady area and loosen or remove clothing if the situation permits. Start cooling him immediately. Immerse him in water and fan him. Massage his extremities and skin and elevate his legs. If conscious, have the soldier/marine slowly drink one full canteen of water.

DANGER

Heatstroke is a medical emergency. Seek medical attention immediately.

WATER SUPPLY

Maintaining safe, clean, water supplies is critical. The best containers for small quantities of water (5 gallons) are plastic water cans or coolers. Water in plastic cans will be good for up to 72 hours; storage in metal containers is safe only for 24 hours. Water trailers, if kept cool, will keep water fresh up to five days. If the air temperature exceeds 100 degrees Fahrenheit, the water temperature must be monitored. When the temperature exceeds 92 degrees Fahrenheit, the water should be changed, as bacteria will multiply. If the water is not changed the water can become a source of sickness, such as diarrhea. Ice in containers keeps water cool. If ice is put in water trailers, the ice must be removed prior to moving the trailer to prevent damage to the inner lining of the trailer.

Potable drinking water is the single most important need in the desert. Ensure nonpotable water is never mistaken for drinking water. Water that is not fit to drink but is not otherwise dangerous (it may be merely oversalinated) may be used to aid cooling. It can be used to wet clothing, for example, so the body does not use too much of its internal store of water.

Use only government-issued water containers for drinking water. Carry enough water on a vehicle to last the crew until the next planned resupply. It is wise to provide a small reserve. Carry water containers in positions that-

- Prevent vibration by clamping them firmly to the vehicle body.
- Are in the shade and benefit from an air draft.
- Are protected from puncture by shell splinters.
- Are easily dismounted in case of vehicle evacuation.

Troops must be trained not to waste water. Water that has been used for washing socks, for example, is perfectly adequate for a vehicle cooling system.

Obtain drinking water only from approved sources to avoid disease or water that may have been deliberately polluted. Be careful to guard against pollution of water sources. If rationing is in effect, water should be issued under the close supervision of officers and noncommissioned officers.

Humans cannot perform to maximum efficiency on a decreased water intake. An acclimatized soldier/marine will need as much (if not more) water as the nonacclimatized soldier/marine, as he sweats more readily. If the ration water is not sufficient, there is no alternative but to reduce physical activity or restrict it to the cooler parts of the day.

In very hot conditions it is better to drink smaller quantities of water often rather than large quantities occasionally. Drinking large quantities causes excessive sweating and may induce heat cramps. Use of alcohol lessens resistance to heat due to its dehydrating effect. As activities increase or conditions become more severe, increase water intake accordingly.

The optimum water drinking temperature is between 10 degrees Celsius and 15.5 degrees Celsius (50-60 degrees Fahrenheit). Use lister bags or even wet cloth around metal containers to help cool water.

Units performing heavy activities on a sustained basis, such as a forced march or digging in, at 80 degrees wet bulb globe temperature index, may require more than 3 gallons of drinking water per man. Any increase in the heat stress will increase this need. In high temperatures, the average soldier/marine will require 9 quarts of water per day to survive, but 5 gallons are recommended. Details on water consumption and planning factors are contained in Appendix G.

HINT: Taking a military issue wool sock, soaking it in water, and wrapping it around a canteen or placing it on a **water** bottle, will cool the water through evaporation.

While working in high desert temperatures, a man at rest may lose as much as a pint of water per hour from sweating. In very high temperatures and low humidity, sweating is not noticeable as it evaporates so fast the skin will appear dry. Whenever possible, sweat should be retained on the skin to improve the cooling process; however, the only way to do this is to avoid direct sun on the skin. This is the most important reason why desert troops must remain fully clothed. If a soldier/marine is working, his water loss through sweating (and subsequent requirement for replenishment) increases in proportion to the amount of work done (movement). Troops will not always drink their required amount of liquid readily and will need to be encouraged or ordered to drink more than they think is necessary as the sensation of thirst is not felt until there is a body deficit of 1 to 2 quarts of water. This is particularly true during the period of acclimatization. Packets of artificial fruit flavoring encourages consumption due to the variety of pleasant tastes.

All unit leaders must understand the critical importance of maintaining the proper hydration status. Almost any contingency of military operations will act to interfere with the maintenance of hydration. Urine provides the best indicator of proper hydration. The following are considerations for proper hydration during desert operations:

- Water is the key to your health and survival. Drink before you become thirsty and drink often, When you become thirsty you will be about a "quart and a half low".
- Carry as much water as possible when away from approved sources of drinking water. Man can live longer without food than without water.
- Drink before you work; carry water in your belly, do not "save" it in your canteen. Learn to drink a quart or more of water at one time and drink frequently to replace sweat losses.
- Ensure troops have at least one canteen of water in reserve, and know where and when water resupply will be available.
- Carbohydrate/electrolyte beverages (e.g., Gatorade) are not required, and if used, should not be the only source of water. They are too concentrated to be used alone. Many athletes prefer to dilute these 1:1 with water. Gaseous drinks, sodas, beer, and milk are not good substitutes for water because of their dehydrating effects.
- If urine is more colored than diluted lemonade, or the last urination cannot be remembered, there is probably insufficient water intake. Collect urine samples in field expedient containers and spot check the color as a guide to ensuring proper hydration. Very dark urine warns of dehydration. Soldiers/marines should observe their own urine, and use the buddy system to watch for signs of dehydration in others.
- Diseases, especially diarrheal diseases, will complicate and often prevent maintenance of proper hydration.

Salt, in correct proportions, is vital to the human body; however, the more a man sweats, the more salt he loses. The issue ration has enough salt for a soldier/marine drinking up to 4 quarts of water per day. Unacclimatized troops need additional salt during their first few days of exposure and all soldiers/marines need additional salt when sweating heavily. If the water demand to balance sweat loss rises, extra salt must be taken under medical direction. Salt, in excess of body requirements, may cause increased thirst and a feeling of sickness, and can be dangerous. Water must be tested before adding salt as some sources are already saline, especially those close to the sea.

COLD

The desert can be combine to produce dangerously cold. The dry air, wind, and clear sky can bone-chilling discomfort and even injury. The ability of the body to maintain body temperature within a narrow range is as important in the cold as in the heat. Loss of body heat to the environment can lead to cold injury; a genera] lowering of the body temperature can result in hypothermia, and local freezing of body tissues can lead to frostbite. Hypothermia is the major threat from the cold in the desert, but frostbite also occurs.

Troops must have enough clothing and shelter to keep warm. Remember, wood is difficult to find; any that is available is probably already in use. Troops maybe tempted to leave clothing and equipment behind that seems unnecessary (and burdensome) during the heat of the day. Cold-wet injuries (immersion foot or trench foot) may be a problem for dismounted troops operating in the coastal marshes of the Persian Gulf during the winter. Some guidelines to follow when operating in the cold are—

- Anticipate an increased risk of cold-wet injuries if a proposed operation includes lowland or marshes. Prolonged exposure of the feet in cold water causes immersion foot injury, which is completely disabling.
- Check the weather-know what conditions you will be confronting. The daytime temperature is no guide to the nighttime temperature; 90-degree-Fahrenheit days can turn into 30-degree-Fahrenheit nights.
- The effects of the wind on the perception of cold is well known. Windchill charts contained in FM 21-10 allow estimation of the combined cooling power of air temperature and wind speed compared to the effects of an equally cooling still-air temperature.

CLOTHING

Uniforms should be worn to protect against sunlight and wind. Wear the uniform loosely. Use hats, goggles, and sunscreen. Standard lightweight clothing is suitable for desert operations but should be camouflaged in desert colors, not green. Wear nonstarched long-sleeved shirts, and full-length trousers tucked into combat boots. Wear a scarf or triangular bandanna loosely around the neck (as a sweat rag) to protect the face and neck during sandstorms against the sand and the sun. In extremely hot and dry conditions a wet sweat rag worn loosely around the neck will assist in body cooling.

Combat boots wear out quickly in desert terrain, especially if the terrain is rocky. The leather drys out and cracks unless a nongreasy mixture such as saddle soap is applied. Covering the ventilation holes on jungle boots with glue or epoxies prevents excessive sand from entering the boots. Although difficult to do, keep clothing relatively clean by washing in any surplus water that is available. When water is not available, air and sun clothing to help kill bacteria.

Change socks when they become wet. Prolonged wear of wet socks can lead to foot injury. Although dry desert air promotes evaporation of water from exposed clothing and may actually promote cooling, sweat tends to accumulate in boots.

Soldier/marines may tend to stay in thin clothing until too late in the desert day and become susceptible to chills--so respiratory infections may be common. Personnel should gradually add layers of clothing at night (such as sweaters), and gradually remove them in the morning. Where the danger of cold weather injury exists in the desert, commanders must guard against attempts by inexperienced troops to discard cold weather clothing during the heat of the day.

Compared to the desert battle dress uniform (DBDU) the relative impermeability of the battle dress overgarment (BDO) reduces evaporative cooling capacity. Wearing underwear and the complete DBDU, with sleeves rolled down and under the chemical protective garment, provides additional protection against chemical poisoning. However, this also increases the likelihood of heat stress casualties.

HYGIENE AND SANITATION

Personal hygiene is absolutely critical to sustaining physical fitness. Take every opportunity to wash. Poor personal hygiene and lack of attention to siting of latrines cause more casualties than actual combat. Field Marshal Rommel lost over 28,400 soldiers of his Afrika Corps to disease in 1942. During the desert campaigns of 1942, for every one combat injury, there were three hospitalized for disease. Hygiene and sanitation are covered in detail in FM 21-10. This section highlights some of the points that are of special importance to the commander in the desert.

Proper standards of personal hygiene must be maintained not only as a deterrent to disease but as a reinforcement to discipline and morale. If water is available, shave and bathe daily. Cleaning the areas of the body that sweat heavily is especially important; change underwear and socks frequently, and use foot powder often. Units deployed in remote desert areas must have a means of cutting hair therefore, barber kits should be maintained and inventoried prior to any deployment. If sufficient water is not available, troops should clean themselves with sponge baths, solution-impregnated pads, a damp rag, or even a dry, clean cloth. Ensure that waste water is disposed of in an approved area to prevent insect infestation. If sufficient water is not available for washing, a field expedient alternative is powder baths, that is, using talcum or baby powder to dry bathe.

Check troops for any sign of injury, no matter how slight, as dirt or insects can cause infection in minor cuts and scratches. Small quantities of disinfectant in washing water reduces the chance of infection. Minor sickness can have serious effects in the desert. Prickly heat for example, upsets the sweating mechanism and diarrhea increases water loss, making the soldier/marine more prone to heat illnesses. The buddy system helps to ensure that prompt attention is given to these problems before they incapacitate individuals.

Intestinal diseases can easily increase in the desert. Proper mess sanitation is essential. Site latrines well away and downwind of troop areas and lagers. Trench-type latrines should be used where the soil is suitable but must be dug deeply, as shallow latrines become exposed in areas of shifting sand. Funnels dug into a sump work well as urinals. Layer the bottom of slit trenches with lime and cover the top prior to being filled in. Ensure lime is available after each use of the latrine. Flies area perpetual source of irritation and carry infections. Only good sanitation can keep the fly problem to a minimum. Avoid all local tribe camps since they are frequently a source of disease and vermin.

DESERT SICKNESS

Diseases common to the desert include plague, typhus, malaria, dengue fever, dysentery, cholera, and typhoid. Diseases which adversely impact hydration, such as those which include nausea, vomiting, and diarrhea among their symptoms, can act to dramatically increase the risk of heat (and cold) illness or injury. Infectious diseases can result in a fever; this may make it difficult to diagnose heat illness. Occurrences of heat illness in troops suffering from other diseases complicate recovery from both ailments.

Many native desert animals and plants are hazardous. In addition to injuries as a result of bites, these natural inhabitants of the desert can be a source of infectious diseases.

Many desert plants and shrubs have a toxic resin that can cause blisters, or spines that can cause infection. Consider milky sap, all red beans, and smoke from burning oleander shrubs, poisonous. Poisonous snakes, scorpions, and spiders are common in all deserts. Coastal waters of the Persian Gulf contain hazardous marine animals including sea snakes, poisonous jellyfish, and sea urchins.

Skin diseases can result from polluted water so untreated water should not be used for washing clothes; although it can be used for vehicle cooling systems or vehicle decontamination.

The excessive sweating common in hot climates brings on prickly heat and some forms of fungus infections of the skin. The higher the humidity, the greater the possibility of their occurrence. Although many deserts are not humid, there are exceptions, and these ailments are common to humid conditions.

The following are additional health-related considerations when operating in a desert environment:

- The most common and significant diseases in deserts include diarrheal and insectborne febrile (i.e., fever causing) illnesses-both types of these diseases are preventable.
- Most diarrheal diseases result from ingestion of water or food contaminated with feces. Flies, mosquitoes, and other insects carry fever-causing illnesses such as malaria, sand fly fever, dengue (fever with severe pain in the joints), typhus, and tick fevers.
- There are no safe natural water sources in the desert. Standing water is usually infectious or too brackish to be safe for consumption. Units and troops must always know where and how to get safe drinking water.
- Avoid brackish water (i.e., salty). It, like sea water, increases thirst; it also dehydrates the soldier/marine faster than were no water consumed. Brackish water is common even in public water supplies, Iodine tablets only kill germs, they do not reduce brackishness.
- Water supplies with insufficient chlorine residuals, native food and drink, and ice from all sources are common sources of infective organisms.

PREVENTIVE MEASURES

Both diarrheal and insectborne diseases can be prevented through a strategy which breaks the chain of transmission from infected sources to susceptible soldiers/marines by effectively applying the preventive measures contained in FM 21-10. Additional preventive measures are described below:

- Careful storage, handling and purification/preparation of water and food are the keys to prevention of diarrheal disease. Procure all food, water, ice, and beverages from US military approved sources and inspect them routinely.
- Well-cooked foods that are "steaming hot" when eaten are generally safe, as are peeled fruits and vegetables.
- Local dairy products and raw leafy vegetables are generally unsafe.
- Consider the food in native markets hazardous. Avoid local food unless approved by medical personnel officials.

- Assume raw ice and native water to be contaminated-raw ice cannot be properly disinfected. Ice has been a major source of illness in all prior conflicts; therefore, use ice only from approved sources.
- If any uncertainty exists concerning the quality of drinking water, troops should disinfect their supplies using approved field-expedient methods (e.g., hypochlorite for lister bags, iodine tablets for canteens, boiling).
- Untreated water used for washing or bathing risks infection.
- Hand washing facilities should be established at both latrines and mess facilities. Particular attention should be given to the cleanliness of hands and fingernails. Dirty hands are the primary means of transmitting disease.
- Dispose of human waste and garbage as specified in FM 21-10. Additional considerations regarding human waste and garbage are—

  - Sanitary disposal is important in preventing the spread of disease from insects, animals, and infected individuals, to healthy soldiers/marines.

  - Construction and maintenance of sanitary latrines are essential.

  - Burning is the best solution for waste.

  - Trench latrines can be used if the ground is suitable, but they must be dug deeply enough so that they are not exposed to shifting sand, and they must have protection against flies and other insects that can use them as breeding places.

  - Food and garbage attract animals-do not sleep where you eat and keep refuse areas away from living areas.
- Survey the unit area for potential animal hazards.
- Shake out boots, clothing, and bedding before using them.

Section III. Environmental Effects on Equipment

Conditions in an arid environment can damage military equipment and facilities. Temperatures and dryness are major causes of equipment failure, and wind action lifts and spreads sand and dust, clogging andjamming anything that has moving parts. Vehicles, aircraft, sensors, and weapons are all affected. Rubber components such as gaskets and seals become brittle, and oil leaks are more frequent. Ten characteristics of the desert environment may adversely affect equipment used in the desert:

- Terrain.
- Heat.
- Winds.
- Dust and sand.
- Humidity.
- Temperature variations.
- Thermal bending.
- Optical path bending.

- Static electricity.
- Radiant light.

The relative importance of each characteristic varies from desert to desert. Humidity, for example, can be discounted in most deserts but is important in the Persian Gulf.

TERRAIN

Terrain varies from nearly flat, with high trafficability, to lava beds and salt marshes with little or no trafficability. Drivers must be well trained in judging terrain over which they are driving so they can select the best method of overcoming the varying conditions they will encounter. Techniques for driving and operating equipment in desert conditions are contained in Appendix C.

Track vehicles are well suited for desert operations. Wheel vehicles may be acceptable as they can go many places that track vehicles can go; however, their lower speed average in poor terrain maybe unacceptable during some operations. Vehicles should be equipped with extra fan belts, tires, (and other items apt to malfunction), tow ropes (if not equipped with a winch), extra water cans, and desert camouflage nets. Air-recognition panels, signal mirrors, and a tarpaulin for crew sun protection are very useful. Wheel vehicles should also carry mats, or channels as appropriate, to assist in freeing mired vehicles.

The harsh environment requires a very high standard of maintenance, which may have to be performed well away from specialized support personnel. Operators must be fully trained in operating and maintaining their equipment. Some types of terrain can have a severe effect on suspension and transmission systems, especially those of wheel vehicles. Tanks are prone to throw tracks when traveling over rocks.

Track components require special care in the desert. Grit, heat, and bad track tension accelerate track failure. Sprockets wear out quickly in sandy conditions. Track pins break more easily in high temperatures and high temperatures also increase rubber/metal separation on road wheels. Proper track tension is critical, as loose track is easily thrown and excessive tension causes undue stress on track components.

Increase the unit PLL of tires and tracks as sand temperatures of 165 degrees Fahrenheit are extremely detrimental to rubber, and weaken resistance to sharp rocks and plant spines, Items affected by mileage such as wheels, steering, track wedge bolts and sprocket nuts, and transmission shafts, must be checked for undue wear when conducting before-, during-, and after-operations maintenance.

HEAT

Vehicle coding and lubrication systems are interdependent. A malfunction by one will rapidly place the other system under severe strain. In temperature extremes, all types of engines are apt to operate above optimum temperatures, leading to excessive wear, or leaking oil seals in the power packs, and ultimately, engine failure. Commanders should be aware which types of vehicles are prone to excessive overheating, and ensure extra care is applied to their maintenance. The following are considerations for ensuring engines do not overheat:

- Check oil levels frequently to ensure proper levels are maintained (too high may be as bad as too low), that seals are not leaking, and oil consumption is not higher than normal.

- Keep radiators and air flow areas around engines clean and free of debris and other obstructions.
- Fit water-cooled engines with condensers to avoid steam escaping through the overflow pipe.
- Cooling hoses must be kept tight (a drip a second loses 7 gallons of fluid in 24 hours).
- Operators should not remove hood side panels from engine compartments while the engine is running as this causes air turbulence and leads to ineffective cooling.

Batteries do not hold their charge efficiently in intense heat. Check them twice daily. The following are additional considerations for maintaining batteries in intense heat:

- Change battery specific gravity to adjust to the desert environment (see vehicle TMs for details).
- Keep batteries full, but not overfilled, and carry a reserve of distilled water.
- Keep air vents clean, or vapors may build up pressure and cause the battery to explode.
- Set voltage regulators as low as practical.
- Increase dry battery supplies to offset high attrition rate caused by heat exposure.

Severe heat increases pressure in closed pressurized systems such as the M2 fire burner unit, and increases the volume of liquids. Ensure that the working pressure of all equipment is within safety limits and be careful when removing items such as filler caps.

Treat Halon fire extinguishers with care. High temperatures may cause them to discharge spontaneously. Put wet rags on them during the hottest part of the day to keep them cooler.

Some items of equipment are fitted with thermal cutouts that open circuit breakers whenever equipment begins to overheat. Overheating is often caused by high ambient temperatures, and can be partly avoided by keeping the item in the shade or wrapping it in a wet cloth to maintain a lower temperature by evaporation.

Flying time and performance of helicopters are degraded as the altitude and heat increases. Helicopter performance is also affected by humidity. Aircraft canopies have been known to bubble under direct heat and should be covered when not in use.

Ammunition must be out of direct heat and sunlight. Use camouflage nets and tarpaulins to provide cover. Ammunition cool enough to be held by bare hands is safe to fire.

Wood shrinks in a high-temperature, low-humidity environment. Equipment, such as axes carried on track vehicles, can become safety hazards as heads are likely to fly off shrunken handles.

Radiators require special attention. Proper cooling-system operation is critical in high-temperature environments. Check cooling systems for serviceability prior to deployment. Local water maybe high in mineral content which will calcify in cooling systems. Distilled water is better since tap water contains chemicals that will form a crusty coating inside the radiator and will ultimately clog it. A mixture of 40 percent antifreeze and 60 percent water is usually acceptable-check your appropriate technical manual to be certain.

During movement, and at operation sites where extremely hot temperatures exist, continuous protection from the heat is necessary for medical items and supplies, which deteriorate rapidly.

Air and all fluids expand and contract according to temperature. If tires are inflated to correct pressure during the cool of night, they may burst during the heat of day. If fuel tanks are filled to the brim at night, they will overflow at midday. Servicing these items during the heat of day can result in low tire pressure, overheating of tires, and a lack of endurance if the fuel tanks were not filled to their correct levels. Air pressure in tires must be checked when the equipment is operating at efficient working temperatures, and fuel tanks must be filled to their correct capacity as defined in the appropriate technical manual. These items should be checked several times a day and again at night.

The major problem with radios in a desert environment is overheating. The following steps can help prevent overheating of radios:

- Keep radios out of direct sunlight.
- Place a piece of wood on top of the radio. Leaving space between the wood and the top of the radio will help cool the equipment. Operating on low power whenever possible will also help.
- Place wet rags on top of radios to help keep them cool and operational. Do not cover the vents.

Any oil or fuel blown onto a cooler (heat exchanger) will gather and quickly degrade cooling. Fix even slight leaks promptly. Do not remove cooling ducts or shrouds. Check them for complete coverage-use tape to seal cracks. Do not remove serviceable thermostats if overheating occurs.

WINDS

Desert winds, by their velocity alone, can be very destructive to large and relatively light materiel such as aircraft, tentage, and antenna systems. To minimize the possibility of wind damage, materiel should be sited to benefit from wind protection and should be firmly picketed to the ground.

DUST AND SAND

Dust and sand are probably the greatest danger to the efficient functioning of equipment in the desert. It is almost impossible to avoid particles settling on moving parts and acting as an abrasive. Sand mixed with oil forms an abrasive paste.

Lubricants must be of the correct viscosity for the temperature and kept to the recommended absolute minimum in the case of exposed or semiexposed moving parts. Lubrication fittings are critical items and should be checked frequently. If they are missing, sand will enter the housing causing bearing failure. Teflon bearings require constant inspection to ensure that the coating is not being eroded.

Proper lubrication is crucial for success. Wipe off all grease fittings before you attach the grease gun and after use. Keep cans of grease covered to prevent sand contamination. Preserve opened grease containers by covering and sealing with plastic bags. Use of grease cartridges in lieu of bulk grease is preferred. All POL dispensing tools must be stored in protected areas to prevent contamination. Place a tarpaulin, or other material, under equipment being repaired to prevent tools and components from being lost in the sand. The automotive-artillery grease possesses a significantly high-temperature capability. If not available, an alternative is general purpose wide-temperature range (WTR) aircraft grease.

Oil should be changed about twice as often under desert conditions as under US or European conditions, not only because grit accumulates in the oil pan, but also because noncombusted low-octane fuel seeps down the cylinder walls and dilutes the reservoir. Diluted lubricants cool less effectively, and evaporate at the higher temperatures generated during engine operation. Oil changes and lubrication of undercarriage points at more frequent intervals will prolong engine and vehicle life under desert conditions. Units employed in desert environments should reevaluate their engine oil requirements and plan accordingly.

Keeping sand out of maintenance areas is critical due to the strong possibility of sand or dust entering the cylinders or other moving parts when the equipment is stripped. Baggies, cloth, or plastic can be used to protect open or disassembled components from blowing sand and dust. The same applies for disconnected water, oil, or other fluid lines. Be sure to cover both ends of the connection if stored. It is essential to have screens against blowing sand (which also provides shade for mechanics). The surrounding ground may be soaked in used oil or covered with rocks to bind it down. Mechanics must keep their tools clean.

Dust and sand can easily cause failure of such items as radio and signal distribution panels, and circuit breakers, and cause small electrical motors to burn out. Wheel and flight control bearings may require daily cleaning and repacking, and engines should be flushed of contaminants daily.

Rotor heads have reduced life, requiring more frequent inspections than in temperate climates. Pay particular attention to sand-caused wear on rotor heads, leading edges of rotor blades, and exposed flight controls. Over 200 pounds of dirt has been known to accumulate in the fuselage area of helicopters operating in desert conditions. These areas must be routinely checked and cleaned to prevent a pound-for-pound reduction in aircraft-lift capability.

Filters must be used when refueling any type of vehicle, and the gap between the nozzle and the fuel tank filler pipe must be kept covered. It takes comparatively little dirt to block a fuel line. Fuel filters will require more frequent cleaning and will need to be checked and replaced often. Engine oil should be changed more often and oil filters replaced more frequently than in temperate climates.

Compression-ignition engines depend on clean air; therefore, examine and clean air cleaners of every type of equipment at frequent intervals. The exact interval depends on the operating conditions, but as a minimum, should be checked at least daily.

Air compressors are valuable pieces of equipment in the desert. They are essential for cleaning air filters and removing dust and sand from components. Intake filters require cleaning daily.

Windblown sand and grit, in addition to heat, will damage electrical wire insulation over a period of time. All cables that are likely to be damaged should be protected with tape before insulation becomes worn. Sand will also find its way into parts of items such as "spaghetti cord" plugs, either preventing electrical contact or making it impossible to join the plugs together. Use a brush, such as an old toothbrush, to brush out such items before they are joined. Electrical tape placed over the ends of spaghetti cords also works.

Radio is the primary means of communications in the desert. It can be employed effectively in desert climates and terrain to provide the reliable communications demanded by widely dispersed forces. However, desert terrain provides poor electrical ground, and a counterpoise (an artificial ground) is needed to improve the range of certain antennas.

Some receiver-transmitters have ventilating ports and channels that can get clogged with dust. These must be checked regularly and kept clean to prevent overheating. Mobile subscriber equipment may require the deployment of additional radio access units (RAU) AN/VRC-191. These assemblages are the primary link for the mobile subscriber radio telephone terminal (MSRT) AN/VRC-97s which are located down to battalion level. The normal operating range of the receiver-transmitter used with these radios may only be 10 kilometers in the desert.

Dust and sand adversely affect the performance of weapons. Weapons may jam or missiles lock on launching rails due to sand and dust accumulation. Sand- or dust-clogged barrels lead to in-bore detonations. Daily supervised cleaning of weapons is essential. Particular attention should be given to magazines which are often clogged, interrupting the feeding of weapons. Cover missiles on launchers until required for use. To avoid jamming due to the accumulation of sand, the working parts of weapons must have the absolute minimum amount of lubrication. It may even be preferable to have them totally dry, as any damage caused during firing will be less than that produced by the sand/oil abrasive paste. Paintbrushes are among the most useful tools to bring to the desert; they are extremely effective in cleaning weapons and optics.

Take precautions to prevent exposure of floppy disks and computers to dust or sand. Covering them in plastic bags is a technique that has worked for several different units. A number of units have successfully operated PLL computers in inflatable medical NBC shelters (MIS 1). This technique has obvious drawbacks since the shelter was not designed for this; however, until a materiel fix is developed, this sort of innovation may be necessary. Compressed air cans, locally purchased from computer vendors, will facilitate the cleaning of keyboards and other components of computer systems.

All optics are affected by blown sand, which gradually degrade their performance due to small pitting and scratching. It is necessary to guard against buildup of dust on optics, which may not be apparent until the low light optical performance has severely deteriorated. It may be advisable to keep optics covered with some form of cellophane film until operations are about to start, especially if the unit is in a sandstorm. A cover that has no sand on the underside should also be used and must be secured so it cannot vibrate against the wind screen. Both of these measures are equally important to tactical security as sun reflected from these optics will reveal positions.

Sand and dirt can easily accumulate in hull bottoms of armored vehicles. This accumulation, combined with condensation or oil, can cause jamming of control linkages. Sand accumulation at the air-bleeder valve can inhibit heat from escaping from the transmission and result in damage to the transmission. Park tactical wheeled vehicles with the rear facing the wind to keep sand out of the radiator. Tracked vehicles should park to protect the engine compartment (grille doors away from wind) from the same sort of damage. The operator's checks and services increase in importance in this environment.

HUMIDITY

Some deserts are humid. Where this is the case, humidity plus heat encourages rust on bare metal and mold in enclosed spaces such as optics. Bare metal surfaces on equipment not required for immediate use must be kept clean and very lightly lubricated.

Items such as optics must be stored in dry conditions; those in use should be kept where air can circulate around them, and should be purged at frequent intervals. Aircraft must be washed daily, particularly if there is salt in the air, using low-pressure sprays.

TEMPERATURE VARIATIONS

In deserts with relatively high-dew levels and high humidity, overnight condensation can occur wherever surfaces (such as metal exposed to air) are cooler than the air temperature. Condensation can affect such items as optics, fuel lines, and air tanks. Drain fuel lines both at night and in the morning (whenever necessary). Clean optics and weapons frequently. Weapons, even if not lubricated, accumulate sand and dirt due to condensation.

THERMAL BENDING

Weapon systems such as the tank cannon are affected in several ways by the desert. One is thermal bending, which is the uneven heating and cooling of a gun tube due to ambient temperature changes. Modem tanks, like the Ml, have been designed to compensate for these factors. The muzzle reference system (MRS) allows the crew to monitor any loss of gun sight relationship and to comet for any error using the MRS update at regular intervals. Ml-series tanks are equipped with a thermal shroud, allowing for more even heating and cooling of the gun tube. Both factors can greatly reduce the accuracy of a tank weapon system. By boresighting at regular intervals and constant monitoring of the fire control system, the tank crew can maximize its readiness. "Gun tube droop" can be countered using the MRS update at least four times in a 24-hour period: at dawn as part of stand-to; at noon to compensate for gun tube temperature chang: before EENT, for TIS reticle confirmation; and at 0100 hours to compensate for gun tube temperature changes.

OPTICAL PATH BENDING

The apparent illusion of target displacement is commonly called refraction. Under certain light and environmental conditions, the path of light (line of sight) may not appear to travel in a straight line. Refraction may cause problems for tank crews attempting engagements at ranges beyond 1,500 meters. Figure 1-17 shows an example of optical path bending in the desert. Refraction may occur in the following conditions:

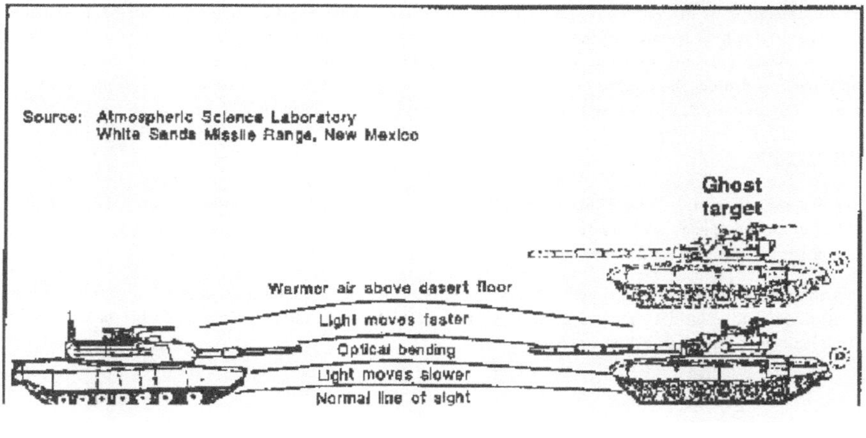

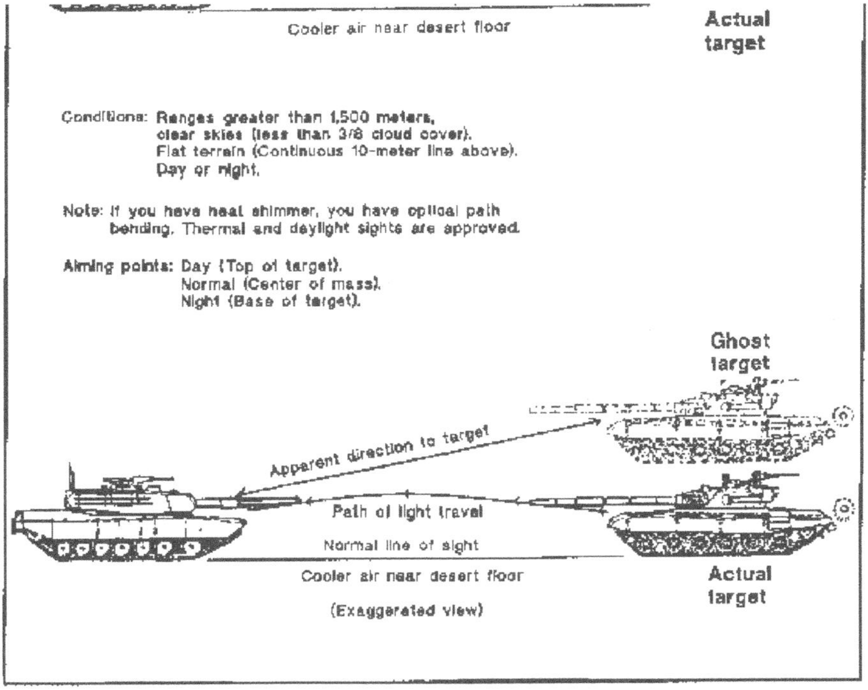

Figure: 1-17. Optical path bending in the desert.

- Day—Clear sky, flat terrain, winds less than 10 miles per hour,
- Night-Clear sky, flat terrain, winds under 4 miles per hour.

NOTE: Any time heat shimmer is present, refraction may also exist.

The effect of refraction is to make the target appear lower during the day; the sight picture, though it appears center of visible mass to the gunner, is actually below the target. This may result in a short round. At night, the effects are the opposite and may result in an over round. Crews must not be fooled by what appears to be a good range from their laser range finder (LRF); the laser beam will refract with other light rays and still hit the desired target.

The most effective measure available to the crew to minimize refraction is an elevated firing position. A position at least 10 meters above intervening terrain will generally negate any effects. If this type of position is not available, a crew operating under conditions favorable to refraction, and having missed with their first round, should apply the following:

- Day--Adjust sight picture up 1/2 target form. See Figures 1-18 and 1-19 for examples of day and night refraction, respectively.
- Night—Adjust sight picture down 1/2 target form.

NOTE: Crews do not need to make a correction for refraction at ranges of less than 1,500 meters.

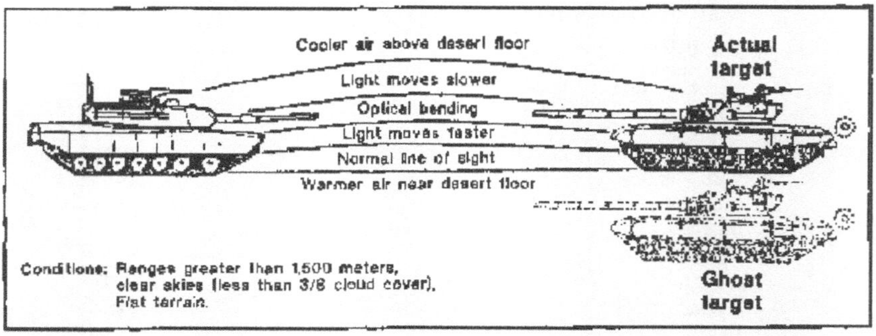

Figure 1-18. Day refraction.

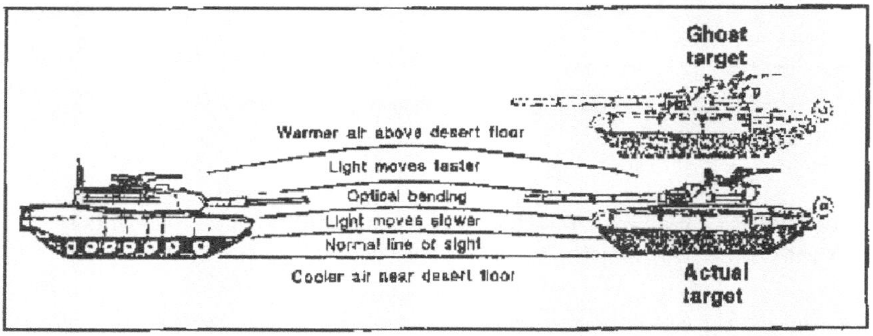

Figure 1-19. Night refraction

STATIC ELECTRICITY

Static electricity is prevalent and poses a danger in the desert. It is caused by atmospheric conditions coupled with an inability to ground out due to dryness of the terrain. It is particularly prevalent with aircraft or vehicles having no conductor contact with the soil. The difference of electrical potential between separate materials may cause an electrical discharge between them when contact is made, and if flammable gases are present, they may explode and cause a fire. Poor grounding conditions aggravate the problem. Be sure to tape all sharp edges (tips) of antennas to reduce wind-caused static electricity. If you are operating from a fixed position, ensure that equipment is properly grounded.

Establish a metal circuit between fuel tankers and vehicles before and during refueling. Ensure the fuel tankers and vehicles are grounded (for example, by a cable and picket or by a crowbar). Grounding of vehicles and equipment should be accomplished in accordance with appropriate operations manuals.

Static electricity is also a hazard with helicopter sling loads. Exercise care when handling and transporting unlike materials that might generate static electricity. Also turn off all switches, uncouple electrical connectors, and ground vehicle or aircraft electrically-operated weapon systems before reaming. Static electricity will also ruin circuit boards and other electronic equipment.

RADIANT LIGHT

Radiant light may be detrimental to plastics, lubricants, pressurized gases, some chemicals, and infrared tracking and guidance systems. Items like CO^2 fire extinguishers, M13 decontamination and reimpregnating kits, and Stinger missiles must be kept out of constant direct sunlight. Optics have been known to discolor under direct sunlight (although this is unusual), so it is wise to minimize their exposure to the sun's rays.

CHAPTER 2

PREPARATIONS FOR DESERT OPERATIONS

This chapter describes preparations for deployment and training for operations in a desert environment. A force sent to a desert theater will fight with the equipment it has in accordance with current doctrine. While equipment and doctrine can be modified to suit the new environment, much will depend on how well soldiers/marines and leaders have mastered their individual training. Units that have trained in Germany and the United States will have the basic technical and tactical skills that can be adapted for desert warfare. Well-trained troops and leaders can adapt quickly to the peculiar conditions of the new environment. If their individual and collective skills have been neglected, no amount of desert lore will remedy the situation.

Section I. Factors to be Considered When Preparing for Desert Operations

When a unit is alerted for operations in a desert environment the commander must first answer some or all of the following questions:

- To what country is the unit going?
- What are the climatic and terrain conditions of that country?
- Will the unit be taking its own equipment overseas?
- What is the tentative timeline for departure? Will there be a period of time where the unit has time for individual training while the vehicles are in transit?
- What unit equipment is being sent overseas and what items will it require for modification (including camouflage painting)?
- What special equipment does the unit require for desert operations?
- What special maintenance is required for weapons and equipment before deployment to or arrival in a desert environment?
- Are there personnel in the unit who—
    - Have desert experience as observers or controllers?
    - Have any experience in desert conditions?
- Are all personnel physically fit?
- How many soldiers/marines are nondeployable?
- What types of operations are expected?

Once these and other questions have been answered the commander must develop a program to bring his unit to a level where it is fully capable of successfully operating in harsh desert conditions. To do this, first set a list of priorities for both individual and unit training. The training priorities listed below are shown as a guide only. They can be modified as necessary depending on the state of readiness of the unit when it is first alerted for desert employment.

Section II. Individual Training

In order to fight and survive in desert operations troops must fully understand the desert environment. The objective of individual training is to prepare the individual for operations in a desert environment. This requires both mental and physical preparation.

To the extent practicable, troops should be acclimatized before arrival in the area of operations. The requirement for acclimatization will vary slightly between individuals, but physical conditioning (fit soldiers/marines acclimatize more easily) is a part of the acclimatization process. Acclimatization should take place in climatic conditions that are similar, or slightly more strenuous, than those of the prospective area of operations.

CAMOUFLAGE AND CONCEALMENT

Camouflage and concealment training may be divided into concealment from the ground and concealment from the air. Particular attention must be paid to movement, color, shadow, and deception. Units should practice erecting and disassembling camouflage netting in order to become more efficient. Well-trained crews can save time and headaches. Camouflage and concealment are equally important for combat service support troops. Appendix E contains information about desert camouflage and concealment techniques.

DRIVER TRAINING

Because of the absence of established roads in desert areas, driving requires experience, individual skill, and physical endurance on the part of the vehicle operator. Driver training exercises should be long and arduous to expose vehicle operators to the rigors of the desert as well as to the effects of fatigue. The need for dispersing and avoiding preceding vehicles is stressed when operating over crusted surfaces or when the trail deteriorates while operating over sand (except suspected mine areas). Training should be directed toward driver proficiency in dune areas, choice of the best ground, selection of proper gear ratios, and driver knowledge and appreciation of the exact capabilities of his vehicle. Driver skill should be developed in taking maximum advantage of momentum, gear shifting, estimating and utilizing proper speeds, and avoiding sudden driving or braking thrusts. Additional driving techniques are contained in Appendix C.

SURVIVAL, EVASION, AND ESCAPE

Convincing a soldier/marine that he is capable of surviving in the desert environment strengthens his Self-confidence, and thus his morale. FM 21-76 contains details on survival, evasion, and escape in the desert, but the following points concerning FM 21-76 should be included in desert survival, evasion, and escape training due to their importance:

- It is unlikely that wells will be poisoned. However, some wells in the North African desert have such strong concentrations of mineral salts that water taken from them may lead to intestinal irritation and subsequent illness.

- Although water is undoubtedly the most important factor in survival, a soldier/marine should not discard his personal weapon or any navigational equipment except in extreme circumstances. Mirrors of any type should be retained for signaling aircraft or other ground forces.

FIELD TRAINING

Following minimum preliminary training in garrison, desert living can only be practiced in the field, often as part of unit training. Important aspects that should be covered include:

- The effects of heat, including possible dehydration and salt loss (the need to maintain the body fluid level).
- The effects of temperature variations.
- The effects of cold weather in the desert.
- First aid for heat illnesses. Each soldier/marine should be issued a memory aid card showing symptoms and immediate treatment.
- Maintenance of morale and the ability of the individual to accept the challenge of the desert. Self discipline and common sense.
- Environmental effects such as those of sand, wind, and light.
- Water discipline.
- Hygiene and sanitation.
- Comet clothing equipment, including how to wear and maintain and clothing.
- Precautions against snakes, scorpions, insects, and disease-bearing organisms.

To the extent possible, the commander should train his unit in terrain and environmental conditions similar to what he expects to find in the operational area. It would be both shortsighted and dangerous for example, to allow water for bathing if the expected operational area is totally waterless. To further accustom the troops to hardships, contact with garrison or other urban areas should be kept to the minimum except for medical or welfare reasons. Once field training has started, necessary supplies should be brought to field locations and items that are unlikely to be available in the operational area (commercial soft drinks and foods) should not be permitted. To gain the maximum value from this training, the unit's exposure to outside influences should be kept to a minimum.

ENEMY ORGANIZATIONS AND TACTICS

This can be taught in garrison on sand tables and map maneuvers, followed by tactical exercises without troops (TINT), and unit exercises in the field. If enemy equipment is available it should be brought to the unit so it can be studied firsthand.

DESERT NAVIGATION

Troops must be thoroughly briefed on the type of terrain and the general environment they will encounter, including-

- Water sources, if any.

- Landmarks or significant permanent terrain features.
- Friendly and enemy areas of operation.
- Prevailing winds.
- Whether or not the local populace is pro or con the US.

This information will assist navigation by reconnaissance units or individuals who become separated from their units. Although maps are the most obvious navigation aids, numerous types of equipment and techniques are available to assist soldiers/marines during desert operations.

Maps

Although maps used in field training will be those of the local area, sufficient maps of the operational area should be obtained to allow distribution for study and possible use during garrison training. This is particularly important if the operational maps use foreign words to describe terrain, such as sebhka, summan, hidiba, and dikaka.

In addition, the grid system on some maps differs from the universal transverse mercator grid system on US maps. In many Middle East countries that were previously under British influence, for example, the Palestine grid system is used on military maps. These maps, generally last surveyed during World War II or the following decade, are widely used, not only in the area of Palestine, but also in Egypt and much of Saudi Arabia. And since they are commonly produced in either 1:100,000 or 1:50,000 scale, they do not mesh with standard US maps. In some instances, accurate maps may not be available. An alternative is to draw the grid lines on attached blank sheets of paper. This method can be highly effective when used in conjunction with navigational aids such as the GPS and LORAN.

Latitude and Longitude Conversions

One of the oldest systematic methods of location is based upon the geographic coordinate system. By drawing a set of east-west rings around the globe (parallel to the equator), and a set of north-south rings crossing the equator at right angles and converging at the poles, a network of reference lines is formed from which any point on the earth's surface can be located. The distance of a point north or south of the equator is known as its latitude. The rings around the earth parallel to the equator are called parallels of latitude or simply parallels. Lines of latitude run east-west but north-south distances are measured between them. A second set of rings around the globe at right angles to the lines of latitude and passing through the poles are known as meridians of longitude or simply meridians. One meridian is designated as the prime meridian. The prime meridian of the system we use runs through Greenwich, England, and is known as the Greenwich meridian. The distance east or west of a prime meridian to a point is known as its longitude. Lines of longitude (meridians) run north-south but east-west distances are measured between them.

Geographic coordinates are expressed in angular measurement. Each circle is divided into 360 degrees, each degree into 60 minutes and each minute into 60 seconds. At any point on the earth, the ground distance covered by one degree of longitude is about 111 kilometers (69 miles); one second is equal to about 30 meters (100 feet). The ground distance covered by one degree of longitude at the equator is also about 111 kilometers, but decreases as one moves north or south, until it becomes zero at the poles. For example, one second of longitude represents about 30 meters at the equator but at the latitude of

Washington, D. C, one second of longitude is approximately 24 meters.

Geographic coordinates appear on all standard military maps, and on some they may be the only method of locating and referencing the location of a point. The four lines that enclose the body of the map (neatlines) are lines of latitude and longitude. Their values are given in degrees and minutes at each of the four comers. In addition to the latitude and longitude given for the four corners, there are, at regular intervals along the sides of the map, small tick marks extending into the the body of the map. Each of these tick marks is identified by its latitude and longitude value. Different methods exist for converting longitude/ latitudes to the military grid system. Special equipment such as the global positioning systems have the capability to convert longitude and latitudes to grid coordinates, or this may be accomplished through manual means outlined in FM 21-26. For more information on map reading and land navigation techniques consult FM 21-26.

Navigation Aids

Navigation aids vary in sophistication and complexity and may include the following:

Sun Compasses/Sextants

These systems can be used on moving vehicles and require accurate timekeeping.

Lensatic Compasses

Individual compass error and local deviation must be known before using the lensatic compass. The lensatic compass cannot be used with any accuracy on dense steel vehicles such as tanks. A crew member should dismount to obtain an azimuth. It is unreliable near large quantities of metal, and can also be affected by underground mineral deposits. Power lines also adversely affect the lensatic compass.

Gyro Compass

An efficient gun azimuth stabilizer (a gyroscope) used on fairly flat ground is useful for maintaining direction.

Fires

Planned tracer fire assists in maintaining bearings, and field artillery and mortar concentrations, preferably smoke (or illumination at night), are useful checks on estimated locations.

Distance Recorders

It is essential to record distance moved, which may be done by using a vehicle odometer.

Beacons

These are particularly useful for aircraft navigation, but can also permit the enemy to locate friendly forces. It may be necessary to place them in open desert with unit locations being marked at certain distances and bearings from them.

Radars

Provided the position of a radar is known, it can measure range and bearings and, therefore, the position of a vehicle.

Aerial Photographs

The advantage of aerial photographs, particularly to aviators, is their ability to show up-to-date views of the variations in color and texture of the desert soil.

Global Positioning Systems

The GPS is a space-based, radio-positioning navigation system that provides accurate passive position, speed, distance, and bearing of other locations to suitably equipped users. The system assists the user in performing such missions as siting, surveying, tactical reconnaissance, artillery forward observing, close air support, general navigation, maneuver, and ground-based forward air control. It can be operated in all types of weather, day or night, anywhere in the world; it may also be used during nuclear, biological, and chemical warfare. It is important to remember these types of devices are aids to navigation; therefore, users should continuously plot their positions. In the event of a GPS failure, you can revert to more traditional navigation and position determination methods.

Position and Azimuth Determining System

The PADS is a highly mobile, self-contained, passive, all-weather, survey-accurate position/navigation instrument used by field artillery and air defense artillery units for fire support missions. The system provides roil-time, three-dimensional coordinates in meters, and a grid azimuth in roils. It also gives direction and altitude. The PADS can be used by the land navigator to assist in giving accurate azimuth and distance between locations. A unit requiring accurate information as to its present location can use PADS to provide this information.

Position Location Reporting System

The position location reporting system (PLRS)/joint tactical information distribution system (JTIDS), hybrid (PJH), is a computer-based system. It provides near real time, secure data communications, identification, navigation, position location, and automatic reporting to support the need of commanders for information on the location, identification, and movement of friendly forces. The PLRS is based on synchronized radio transmissions in a network of users controlled by a master station. The major elements of a PLRS community include the airborne, surface vehicular, and man-pack users; the PLRS master station; and an alternate master station. The system can handle 370 user units in a division-size deployment per master station with a typical location accuracy at 15 meters. The man-pack unit weighs 23 pounds and includes the basic user unit, user readout, antenna, backpack, and two batteries.

Dead Reckoning

The simplest system of navigation is known as dead reckoning. This is a means of finding where an individual is located by a continuous plotting of where he has been. More exactly, dead reckoning consists of recording and plotting a series of courses, each measured as to the distance and direction horn a known point, to provide a plot from which the position can be determined at any time. In the desert, the direction traveled is determined with a compass and the distance is measured by counting paces or reading the odometer of a vehicle. Detailed information on navigation by dead reckoning is contained in FM 21-26.

Shadow-Tip Method

This method provides orientation by reading the way the sun casts shadows. To use the shadow-tip method, find a fairly straight stick about 1 meter long and follow these steps:

- Step 1. Push the stick into the ground at a fairly level, brush-free spot where a distinct shadow will be cast. The stick need not be vertical; inclining it to obtain a more convenient shadow, in size or direction, does not impair the accuracy of the shadow-tip method.
- Step 2. Mark the tip of the shadow with a small peg, stick, stone, twig, your finger, hole in the sand, or other means. Wait until the shadow's tip moves a few inches (if you use a 1-meter stick, 10 to 15 minutes should be enough time),
- Step 3. Mark the new position of the shadow's tip.
- Step 4. Draw a straight line from the first mark to the second mark and extend it about a foot past the second mark.
- Step 5. Stand with the toe of the left foot at the first mark and the toe of the right foot past the line you drew.

You are now facing true north. Find other directions by recalling their relation to north. To mark directions on the ground (to orient others), draw a line at right angles to the first line, forming a cross and mark the directions.

If you cannot remember which foot to place on the first rock (see step 5), remember this basic rule for telling east from west: the sun rises in the east andsets in the west (but rarely due east or due west). The shadow's tip moves just the opposite. Therefore, the first shadow-tip mark is always in the west direction and the second mark in the east direction, everywhere on earth. Figure 2-1 depicts finding your direction by using the shadow-tip method.

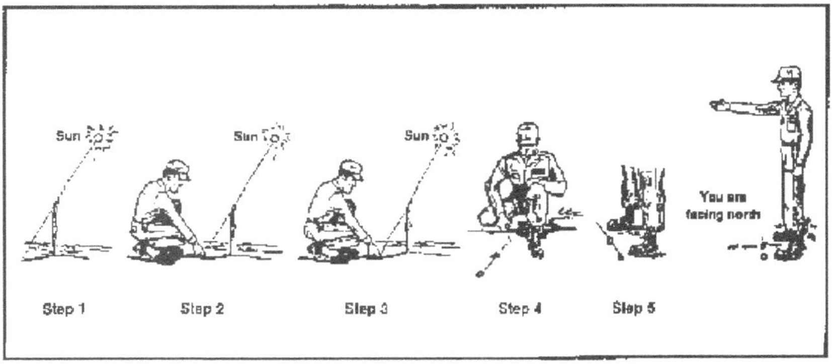

Figure 2-1.Finding direction Using shadow-up method.

Watch/Sun Method

An ordinary analog watch (with hands) can be used to determine the approximate true north in the North and South Temperate Zones. The North Temperate Zone is north of the equator and the South Temperate Zone is south of the equator. The temperate zones extend from latitude 23-1/2 degrees to 66-1/2 degrees in both hemispheres. In the North Temperate Zone only, the hour hand is pointed toward the sun (see Figure 2-2). A north-south line can be found midway between the hour hand and 12 o'clock. This applies to standard time. For daylight savings time, the north-south line is found midway between the hour hand and 1 o'clock. If there is any doubt as to which end of the line is north, remember that the sun is in the north, and remember that the sun is in the eastern part of the sky before noon and in the western part in the afternoon.

The watch may also be used to determine direction in the South Temperate Zone (see Figure 2-2). However, it is used a bit differently. Twelve o'clock is pointed toward the sun, and the north-south line will be halfway between 12 o'clock and the hour hand. If on daylight savings time, the north-south line lies midway between the hour hand and 1 o'clock.

On cloudy days, place a stick at the center of the watch and hold it so that the shadow of the stick falls along the hour hand in the North Temperate Zone. In the South Temperate Zone, the shadow falls along a line from the center of the watch through 12 o'clock. Direction is then determined using the appropriate technique.

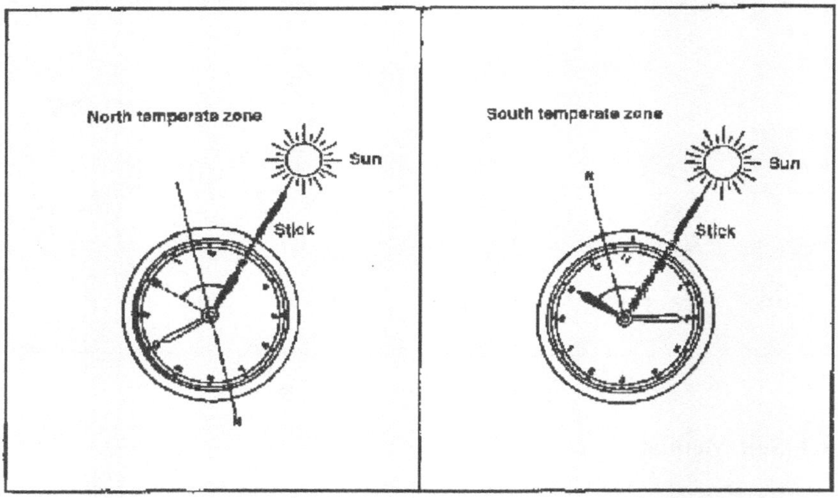

Figure 2-2. Using a watch to find north.

Celestial Navigation Method

North of the equator, locating your direction at night can be determined by locating the North Star. To find the North Star, look for the Big Dipper. The two stars at the end of the bowl are called pointers. In a straight line out from the pointers is the North Star (at about five times the distance between the pointer stars). The Big Dipper rotates slowly around the North Star and does not always appear in the same position (see Figure 2-3).

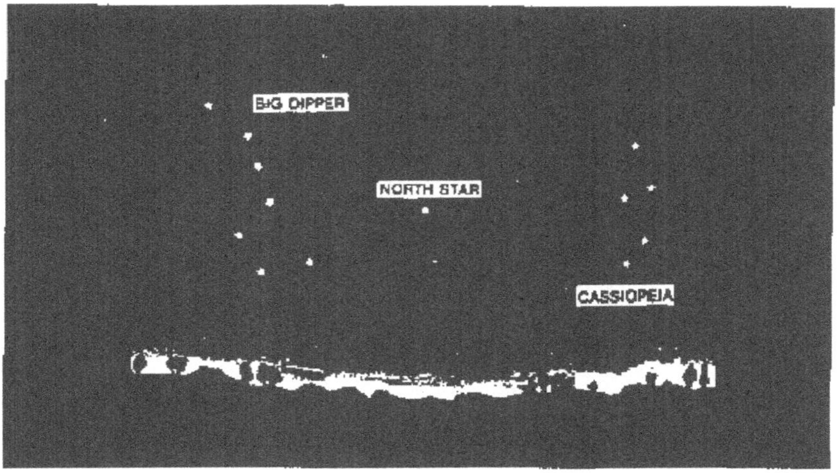

Figure 2-3. ***Big*** Dipper.

The constellation Cassiopeia can also be used. This group of five bright stars is shaped like a lopsided M (or W when it is low in the sky). The North Star is straight out from the center star, about the same distance as from the Big Dipper. Cassiopeia also rotates slowly around the North Star and is always almost directly opposite the Big Dipper. Its position, opposite the Big Dipper, makes it a valuable aid when the Big Dipper is low in the sky, possibly out of sight because of vegetation or high terrain features.

South of the equator, the constellation Southern Cross will help you locate the general direction of the south and, from this base, any other direction. This group of four bright stars is shaped like a cross that is tilted to one side. The two stars forming the long axis, or stern, of the cross are called the pointers. From the foot of the cross, extend the stem five times its length to an imaginary point (see Figure 2-4). This point is the general direction of south. From this point, look straight down to the horizon and select a landmark.

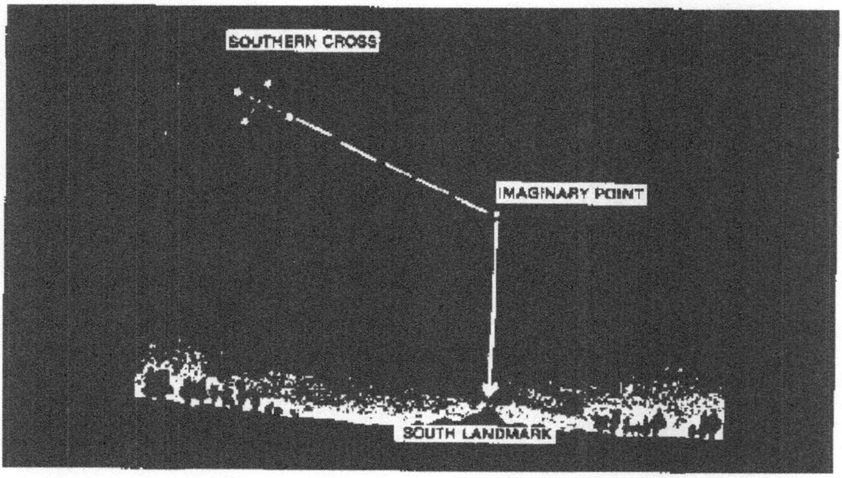

Figure 2-4. Southern Cross.

OPERATIONAL AREA (HOST COUNTRY)

A description of the host country should cover only those facts that apply to forthcoming operations, for example:

- Geographic description.
- Climate (throughout the year).
- Religious issues and constraints.
- Cultural differences and special considerations, important customs, and the behavior expected of US military personnel. (Such as speaking to a woman in some Arab countries, which can be offensive to the local inhabitants.)
- Population density.
- Industry and agriculture.
- Language(s) (phrase books may be issued).
- Communications and transportation network.
- The armed forces (and possibly police), including organization, equipment, and rank structure.
- The situation that has led to the introduction of US forces and reasons why US forces are being introduced. No soldier/marine should have to question why he is fighting for a country other than his own, if this is the case.

Treatment of these subjects will vary in degree according to category. Personnel who may require additional information such as the country's history, can find it in the appropriate DA 550-series pamphlet.

DESERT MANEUVER

Chapter 3 describes the influence of the desert environment on tactical operations. This subject should first be taught to a limited number of leaders and commanders as a theoretical subject, down to platoon level. Leaders should then train their units during unit training. The emphasis should be on small unit tactics, including combined arms operations. Additional subject matter that should be covered includes -

- Terrain in the operational area, emphasizing differences and similarities with the training areas the units will use.
- Application of concealment, using terrain and artificial means such as smoke, and the application of maneuver techniques.
- Mobility in the desert.
- Command and control techniques for desert operations.
- Navigation.
- Conduct of fire in desert operations.
- Resupply during desert operations.
- Special equipment techniques.

GENERAL TRAINING

Nearly all equipment will be affected in one way or another by the environment as described in Chapter L The purpose of this training is to train operators. Training should include-

- Likely effects on the equipment they operate.
- Efficient operations of the equipment within the limits imposed by the environment, including tactical limitations of the equipment, for example, helicopters may have difficulties flying NOE; and radios will normally be operated on reduced output due to the environment and enemy ECM.
- Preventive maintenance-employing any special techniques required by the desert environment. The appropriate equipment technical manual or lubrication order provides specific information concerning hot climate operations and maintenance.
- Basic desert recovery and repair techniques, including defensive measures, and camouflage required during recovery and repair operations.

Instruction must be oriented toward the expected operational area. For example, it is possible to keep radios cool by using ice packs, but if ice packs are not going to be available in the area of operations, then do not teach troops this technique as it will not be practiced.

SPECIAL MAINTENANCE AND SUPPLY TECHNIQUES FOR STAFF AND LEADERS

Special maintenance techniques that need to be addressed are the same as those taught to specialists, however, they only need to emphasize aspects that ordinarily require control or supervision, or affect the employment of equipment in desert terrain. This training should include any special handling techniques required in the operational area. The importance and difficulties of supply in desert operations are described in Chapter 4. Training should be modified according to-

- Modified tables of organization and equipment (MTOE) and mission of the unit.
- Supply situation expected in the area of operations.
- Capabilities of logistic units likely to support unit operations with special attention given to units not normally found in conventional operations, for example, well-drilling teams, refrigeration assets available in the theater, and transportation cargo carrier companies.

NUCLEAR, BIOLOGICAL, AND CHEMICAL (NBC) TRAINING

Wearing protective clothing, flak vests and masks in the desert environment will make a person extremely uncomfortable. Troops should not participate in strenuous activity while wearing protective clothing until they are acclimatized. Training in MOPP gear should become progressively more strenuous. Physical training sessions in field protective masks progressing to foot marches in MOPP 4 is a recommendation. Use of protective clothing in severe desert heat is described in Chapter 1 and Appendix D. Points that should be emphasized during training are-

- The value of being uncomfortable rather than dead.
- The need to avoid heat illness by¬

    - Reducing the labor rate to the minimum, and delaying work until cooler hours.

    - Maintaining proper body water and salt levels, particularly during a time of chemical threat.

    - Detecting the first symptoms of heat illness in others by constant vigilance.

    - Increasing the time factor of an operation as troops will move slowly when wearing protective clothing.

DESERT TERRAIN APPRECIATION

When training soldiers/marines to appreciate desert terrain, leaders should focus on the effects of the different types of desert terrain on the capabilities and limitations of unit equipment. Highlight the impact of the terrain on vehicular trafficability, fields of fire, and observation in the likely operational area. When possible, crews and small unit leaders should learn to appreciate desert terrain from practical experience in terrain as nearly similar as possible to that in the likely combat zone.

MEDICAL TRAINING CONSIDERATIONS

The unit surgeon can provide valuable information on the medical implications of operations in the desert environment. He can advise the unit commanders on measures to take to ensure training includes preventive medicine concepts essential to keeping nonbattle injuries to a minimum. Bum casualties should also be a medical training consideration as these will be the most likely casualties in a mechanized environment. Hydration and mouth-to-mouth resuscitation of injured personnel in a field protective mask should also be a training consideration. Nonbattle casualties, due to a lack of consideration of preventive medicine concepts, can far outnumber combat casualties.

Section III. Unit Training

When determining unit training requirements, the commander must first consider the training level of his unit when alerted for deployment. When time does not permit a comprehensive training program, the commander must concentrate on those areas where his unit is least proficient, considering the priorities previously described. In order to operate in the desert environment, the unit must, above all, be physically fit, so physical conditioning is of paramount importance.

COMBINED ARMS AND SERVICES TEAM

The greatest combat power results when leaders synchronize combat, combat support, and combat service support systems to complement and reinforce one another. The slice concept refers to CS and CSS units task organized to support a particular maneuver or combined arms unit. Leaders should routinely practice habitual relationship and cross attachment of units. Training as combined arms teams is critical to successful desert operations.

PHYSICAL CONDITIONING AND ACCLIMATIZATION

To the extent possible, physical conditioning and acclimatization should take place simultaneously. When a unit is training in a hot environment, begin physical training at night or during the cooler part of the day and work up to rigorous training, such as foot marches in open sand terrain at midday. Emphasis on mounted operations in desert warfare does not imply that foot marching can be totally disregarded. Physical conditioning must be continued after arrival in the area of operations.

Medical advice should always be available during periods of physical training in hot weather. Training several days prior to deployment to combat the effects ofjet lag works well, if performed as a command directive. Avoiding coffee and alcohol, and drinking plenty of water will also assist in overcoming the effects ofjet lag. Planning for the change in time zones and developing a sleep plan are recommended and will facilitate readiness.

WEAPONS TRAINING

Soldiers/marines must train to proficiency at all ranges but accuracy at maximum effective range both in daytime and nighttime must be emphasized. Firing should also be practiced during the heat of day to condition troops to heat haze and mirages. Emphasis must also be placed on maintenance of individual weapons in view of sandy desert conditions.

NBC TRAINING

The purpose of unit NBC training is to train individual soldiers/ marines to become proficient as a team while wearing protective clothing and masks, and when in combat vehicles-while buttoned up. This training should be conducted both in daytime and nighttime, until the unit can operate effectively under these conditions. Additional information and guidance on the effects of the environment on NBC weapons are contained in Appendix D.

MARCH DISCIPLINE

Although of particular importance to combat support and combat service support units, all units should be trained in tactical road marches. Training should emphasize-

- Off-road movement over open terrain. Irregular spacing when moving in convoy.
- The need to maintain sufficient distances between vehicles to preclude "dust blindness."
- Actions to be taken when stuck in sand, and when a vehicle breaks down.
- Vehicle camouflage.
- Removal of tracks which would reveal friendly locations.
- The need for dispersion when halted.
- Air defense drills.
- Practicing incoming artillery drills.

OBSTACLES AND BARRIERS

In some desert areas, natural obstacles such as wadis or other terrain features can be found. Often, however, it will be necessary to use artificial obstacles if enemy movement must be slowed. A minefield, to be of any tactical value in the desert, must usually cover a relatively large area, so mechanical means and engineer support are required. Since there are often too many avenues of approach to be covered with mines, it is usually best to employ mines to cover any gaps between units, especially at night. Minefields are most effective when they can be covered by observation or fire. During unit training, soldiers/marines should be trained to lay mines wearing gloves, since human scent attracts desert animals who may attempt to dig them up. Emphasis should be on antitank minefield since combat vehicles are the most dangerous threat.

SCOUTING, SURVEILLANCE, AND PATROLLING

Effects of the environment on scouting, observation, and surveillance techniques are described in Chapter 3. Effects of the environment on surveillance, target acquisition, and night observation devices are described in Chapter 1.

ADJUSTMENT AND CONDUCT OF FIRE

The principles for adjustment and conduct of fire in the desert are the same as for operations in more temperate climates. However, the following considerations, somewhat peculiar to desert operations, should be kept in mind:

- Obscuration from sand, dust, smoke, or a combination of these, can affect direct-fire adjustment.
- There may be major inaccuracies of initial rounds from indirect-fire weapons due to misjudgment of target location.
- The target may be concealed by sand or dust if rounds land short, on, or near the observer target line.
- Heat haze and mirages can mislead gunners and observers as to target location. This condition can particularly affect antitank guided missile gunners.

Direct-fire gunners may have to depend on flank observers, who may be any individual on the battlefield equipped with a radio. If observation is lost, subsequent corrections are very unlikely to cause a second round hit. The following considerations can help to overcome the obscuration or sending problem:

- An observer requesting indirect fires needs to ensure that initial rounds land beyond the target to preclude short rounds obscuring the target, and then adjust accordingly.
- By remembering the greatest impact of heat haze (which varies throughout the day) is on ATGM gunners-when both gunner and target are within 2-3 feet of the desert surface.

AIR DEFENSE

In desert operations any type of unit, be it tank, infantry, trains, tactical operations centers, or supply points, can expect to be a target for air attack. Air attacks may be from fighter bomber aircraft using cannons, missiles, bombs, napalm, and machine guns, or from attack helicopters using machine guns, rockets, or missiles.

Enemy air superiority should be assumed during all field training and simulated fighter bomber attacks and attack helicopter missions should be flown against the unit whenever possible. When practical, aerial photographs of positions should be taken, and pilots interviewed to assist in the critique of air defense, both passive and active. Points that should be emphasized during training are mentioned in the following paragraphs.

Passive air defense measures should be taken routinely. When stopped for any period of time, take every advantage of whatever cover and concealment are available. As previously described, natural cover and concealment will be difficult to find in many desert areas. Nevertheless, vehicles, particularly unarmored vehicles, should be irregularly dispersed and dug-in, or revetments provided. When appropriate, air guards, trained in aircraft recognition, should be posted, with clear instructions on actions to be taken when aircraft are sighted. Units not being attacked by aircraft, but in close proximity to the attack, may desire to remain stationary in order to avoid detection. Reducing infrared signatures is also a passive anti-air consideration. Artificial camouflage can be used as described in Appendix E.

Active air defense techniques used in desert operations are the same as those described in other doctrinal manuals, appropriate to the level of command. However, at small unit level, additional emphasis should be given to air defense using small arms. When combat vehicles on the move are engaged by enemy aircraft, their immediate action will depend on whether or not they are maneuvering in contact with the enemy. If they are in contact they should continue to maneuver, relying on overwatch elements and air defense artillery to engage attacking aircraft.

Vehicles about to be engaged by enemy aircraft in open desert where cover is not available, should move perpendicular to the attacking aircraft to evade rocket or machine gun fire. Engage the aircraft with small arms fires, if possible. Meanwhile, the remainder of the unit should mass small arms fire to the aircraft's front. Sudden variations in course may also distract the pilot.

COMMUNICATIONS

Good communications in desert terrain will often depend on the state of mind of the operators. They must be enthusiastic, persistent, and determined to make and maintain contact. Unit training should concentrate on ECCM techniques. When conducting field training, higher headquarters can provide assistance in the form of small teams to jam unit nets. Practice actions to be taken when radio contact is lost due to heat (described in Appendix C).

FRATRICIDE CONSIDERATIONS

The following are fratricide considerations when operating in a desert environment and may be topics for fratricide awareness training:

- Because of the absence of easily identifiable terrain features in the desert, knowing your exact location can be especially important. Be in the right place at the right time. Use position location/navigation (GPS) devices; know your location and locations of adjacent units (left, right, leading and follow on). Synchronize tactical movement.
- Ensure positive target identification. Review vehicle/weapons ID cards; know at what ranges and under what conditions positive ID of friendly vehicles/weapons is possible. This is especially important in the desert due to the likelihood of weapons being able to fire at their maximum ranges.
- Maintain situational awareness-be aware of current intelligence, unit locations/dispositions, denial areas (minefields/FASCAM), contaminated areas (e.g., ICM and NBC), SITREPs, and METT-T. This can be more difficult in desert environments because navigation is more difficult.
- Conduct individual and collective (unit) fratricide awareness training. This includes target identification/recognition; fire discipline; and leader training.
- Use common language/vocabulary. Use doctrinally correct, standard terminology and control measures (e.g., fire support coordination line (FSCL), zone of engagement, and restrictive fire line (RFL)).
- Consider the effects of key elements of terrain analysis on fratricide. These include observation and fields of fires, cover and concealment, obstacles and movement, key terrain, and avenues of approach.
- Gun tube orientation can also assist in avoiding/preventing fratricide.

CHAPTER 3

OPERATIONS IN DESERT CONDITIONS

In the two previous chapters the environment and its effects on personnel and equipment and preparations for desert operations were described. This chapter describes desert operations and is divided into three sections.

Section I. How the Desert Environment Affects Tactical Operations

The key to success in desert operations is mobility, and this was clearly evident in ground operations in Desert Storm. The tactics employed to achieve victory over Iraq were wide and rapid flanking movements similar to those Rommel and Montgomery executed in North Africa.

Trafficability and cross-country movement become critical to desert operations when using these tactics. Trafficability is generally good in the desert and cross-country movement is a lesser problem, but not always. Salt marshes can create NO-GO conditions during the rainy season. Sand can also bog down traffic and make foot movement slow and exhausting. The steep slopes of dunes and rocky mountains can make vehicular movement a NO-GO. The wadis create cross-compartmented terrain. The banks of these stream beds can be steep and unconsolidated. Then, when it rains, it becomes a torrent of dangerously rushing water, leading to flat lake beds that can create NO-GO mud conditions. Rock quarries and mining areas can also adversely affect mobility and trafficability. Often these areas are not reflected on maps. Satellite imagery can be helpful in identifying these areas, as was the case in Kuwait during Operation Desert Storm. In rocky terrain, tires can easily be punctured by sharp angular debris; however, overall movement is mostly uninhibited. Given ample fuel, water, and other resources, units can go around natural and man-made obstacles.

Movement can easily be detected because of sand and dust signatures left due to the loose surface material. In an actual engagement, this may not be all that bad because a unit is obscured from direct fire while advancing, but the element of surprise may be lost. Moving at night becomes the logical choice. The dust is still there, and vehicles (which should be widely spaced) can get separated. But at night, reflection of the sun's rays from glass, mirrors, or metal, which can give away movement and positions up to 20 kilometers away, is not a concern.

Using the ability to make fast and wide flanking movements, a unit can encircle and cut off enemy forces. The Israeli forces under General Ariel Sharon did just that to the Egyptian Third Army in the 1973 War, and the British did the same to the Italians in North Africa in January 1941. In Desert Storm, the night-fighting AH-64 helicopters, combined with field artillery fires, made for an unbeatable team in this regard.

Land navigation is a challenge during movement in the wide expanses of many arid regions. There are few landmarks to key on, and maps and even photos can become dated quickly, especially in the sandy deserts where dunes migrate. The global positioning system (GPS) with the small lightweight GPS receivers (SLGRs) is a major aid for desert operations.

Refuel and resupply operations require periods in which forces assume the defense, but only temporarily. The flat sandy desert topography that is characteristic of Saudi Arabia is not conducive to defense, compared to rocky plateau topography. In mountains and canyons, a defensive posture can be favorable. Controlling the passes, as mentioned earlier, can essentially close off vast areas to an attacker and make it extremely costly for him.

While a unit is in the defense, it needs both ground and air reconnaissance to detect enemy movement at long range. Obstacles must be placed in all types of topography, primarily to slow advances and channel columns. Neglecting these security measures in the flat sandy regions can lead to disaster.

MILITARY ASPECTS OF THE TERRAIN

The following paragraphs describe how terrain affects tactical operations in the desert. This discussion follows the outline of the terrain analysis process summarized by the factors of OCOKA.

Observation and Fields of Fire

Observation and fields of fire are generally excellent in most desert areas. The atmosphere is stable and dry, allowing unrestricted view over vast distances, but this can also be a problem. Range estimation by "gut feeling" is subject to error. The effective ranges of weapons can easily be reached, and a correct estimation of maximum ranges is critical for all weapons, especially for wire-guided munitions.

Flat desert terrain permits direct-fire weapons to be used to their maximum range. Open terrain and a predominantly clear atmosphere generally offer excellent long-range visibility; but at certain times of the day visibility may be limited or distorted by heat.

Two primary considerations in the desert environment are longer range observation, and fields of fire at the maximum effective ranges for weapons. However, rapid heating and cooling of the atmosphere hinder these factors and cause distortion of ranges to the aided and unaided eye. Mechanical and electronic means must be used to verify estimated ranges such as GSR and laser range finders. Boresight and zero more frequently at standard ranges.

The desert is not absolutely flat, so weapons are sited to provide mutual support. Dead space is a problem. Even though the landscape appears flat, upon closer inspection it can be undulating with relatively deep wadis and depressions. These areas must be covered by indirect fire.

When on the offense, attacks should be initiated with the sun at or near one's back whenever possible. This eliminates most shadows that degrade optical weapon guidance and makes visual target acquisition difficult.

When there is no usable dominant terrain available, the only means of observation may be from an aeroscout, or limited to short-range observation by the vehicle commander. Other visibility problems are caused by heat distortion. Heat waves at the desert surface cause images to shimmer making positive identification difficult and degrade depth perception. Ranges to targets may be distorted from heat rising from the desert surface. Use range finders to verify correct distances. Be prepared to use bracketing techniques with large adjustments to hit an enemy target with artillery.

Radars are unlikely to be affected by heat haze so they could be valuable on flat terrain during midday heat if optical vision is hopelessly distorted; however, they arc almost useless in sandstorms. Image intensification is of limited value in sandstorms, and depends on the phase of the moon at night. If there is no moon, use artificial illumination outside the field of view of the system.

Since thermal imagery devices depend on the difference between ambient temperature and equipment temperature, they are more useful at night than in the day. Because of the distinct advantages of thermal sights, they should be used as the primary sighting systems for vehicles so equipped.

Correction of field artillery fires, especially those of larger pieces, may be complicated by dust hanging in the air following the impact of ranging rounds. Forward observers should consider placing initial rounds beyond a target rather than short of the target. Observation of fires, especially direct fires by tanks, may be difficult due to dust clouds, so wingmen may have to observe direct fires.

Cover and Concealment

Cover and concealment are generally scarce in the desert. The flat sandy deserts provide little if any natural cover or concealment, especially from aerial attack or reconnaissance. Ground concealment and protection from fire can be found behind dunes or in wadis. Troops must be aware of the potential for flash floods when using wadis for ground concealment.

Some arid regions have vegetation that can provide limited concealment from ground observation. In rocky, mountainous deserts, cover and concealment are best found behind boulders and in crevices. Daytime vehicular movement eliminates nearly any possibility of surprise, as dust trails created by the traffic can be spotted for miles. At night noise and light discipline is critical, as both sound and light travel great distances because of the unobstructed flatness of the terrain and atmospheric stability. Camouflage can be effectively employed to improve on natural cover and concealment. See Appendix E for additional information on concealment and camouflage.

Obstacles

Natural obstacles do exist in the desert, and arid regions are well suited for man-made obstacles. The wadis and steep slopes of escarpments, mountains, hills, and dunes hinder cross-country movement. Sand dunes may stretch for miles and prevent direct movement across their length. These sand dunes are often more than 100 feet in elevation and consist of loose sand with high, steep downwind faces that make vehicular traversing next to impossible. Aerial reconnaissance immediately before any large movement is advisable because sand dunes migrate with shifting winds and they may not be where maps or even photographs show them.

In the Desert Storm area, the salt marshes have a crust on the top that can deceive a vehicle driver. These dry lake beds can become obstacles, especially in the wetter seasons when the water table is higher. A top crust forms on the surface, but below the crust the soil is moist, similar to marsh conditions. The surface may look like it has good trafficability, but the crust will collapse with the weight of a vehicle, and the vehicle becomes mired. The high premium on fuel and time makes it costly to go around these natural obstacles.

Sandy deserts are ideal for employing minefield. Although windstorms can reveal previously buried mines, these mines can still channel movement and deny access to certain areas. The battles of the Bi'R Hacheim Line and El Alamein were influenced by minefield. Other obstacles include ditches, revetments, and barriers, such as the Bar Lev Line along the Suez Canal, made by bulldozing sand mounds or by blasting in rocky mountainous areas to close passes.

Key Terrain

Key terrain in the desert can be any man-made feature, mountain pass, or source of water, and of course, high ground. Because there are few man-made features in the desert, those that do exist can become important, perhaps even key.

Passes through steep topography are also likely to be key, again because they are so few. The North African campaigns of World War II focused on the control of passes, specifically the Solium and Halfaya. In the Sinai Wars between Egypt and Israel, the Mitla, Giddi, and Sudar passes were key. In Afghanistan, control of the mountain passes provided the Mujahideen safe haven from the Soviets. Oases, where wells exist, become important for water resupply. The high ground in desert terrain is usually key terrain. The relative flatness and great distances of some deserts, such as in Iraq, make even large sand dunes dominant features.

Avenues of Approach

Avenues of approach are not clearly defined in arid regions. The vast, relatively flat areas permit maneuver from virtually any direction. This point became obvious to units establishing defensive positions in Desert Storm. Wide envelopments are possible, as demonstrated in the Desert Storm ground campaign. Modem sensor technology, limited natural concealment, and improved observation make the element of surprise a challenge. Yet, surprise was achieved during Desert Storm-Iraqi commanders were shocked when they discovered US tanks in their perimeters.

The major limitation with respect to avenues of approach may be fuel. The great distances a unit must travel to outflank enemy positions require significant amounts of fuel and complicate resupply. In mountainous and canyon topography avenues are much more limited, and the wadis and valleys are likely to be the only possible access routes. Any roads that do exist are probably in the valleys. Nevertheless, none of the considerations outlined above are reasons to disregard flanking movements.

MANEUVER

Army operations are ideally suited to desert environments. Its thrust of securing and retaining the initiative can be optimized in the open terrain associated with the desert environments of the world. In that environment, the terrain aspect of METT-T offers the potential to capitalize on the four basic tenets of the doctrine initiative, agility, depth, and synchronization.

Initiative

Israeli efforts in 1967 and initial Egyptian assaults in 1973 clearly illustrate the effects of initiative in the desert environment.

Agility

The Egyptian success in 1973 was negated by their failure to ensure agility. Conversely, the Israeli actions on the flanks of the Egyptian force demonstrated the effects of a force capable of rapid and bold maneuver.

Depth

Depth does not necessarily relate to distance. In the nonlinear battlefield offered by the desert, depth often equates to an agile reserve force of sufficient size to counter enemy efforts into flanks and rear areas. Depth is also a concept of all-around defense for forces-the ability to fight in any direction.

Synchronization

To a large measure, the German successes against the British in the Western Desert were due to their ability to synchronize their operating systems. More recent events illustrate this tenet between and internal to, operating systems. Heavy/light operations have demonstrated that light forces can be key to achieving tactical and operational momentum. The Israeli airmobile assault against supporting artillery in the 1967 battle of Abu Ageila is a good example of the effective use of light forces in this type of environment.

Maneuver must be at the maximum tactical speed permitted by the terrain, dust conditions, and rate of march of the slowest vehicle, using whatever cover is available. Even a 10-foot sand dune will cover and conceal a combat vehicle. Air defense coverage is always necessary as aircraft can spot movement very easily due to accompanying dust clouds. In some situations movement may be slowed to reduce dust signatures. Rapid movement causes dramatic dust signatures and can reveal tactical movements.

Another consideration during maneuver is dust from NOE flight, which can be seen as far as 30 kilometers. This is especially true when the enemy is stationary. Aeroscouts must use caution to avoid blundering into enemy air defense weapons.

To achieve surprise, maneuver in conditions that preclude observation, such as at night, behind smoke, or during sandstorms. In certain circumstances, there may be no alternative to maneuvering in terrain where the enemy has long-range observation. Then it is necessary to move at the best speed possible while indirect fires are placed on suspected enemy positions. Speed, suppressive fires, close air support, and every other available combat multiplier must be brought to bear on the enemy.

Tactical mobility is the key to successful desert operations. Most deserts permit good to excellent movement by ground troops similar to that of a naval task force at sea. Use of natural obstacles may permit a force to establish a defensive position that theoretically cannot be turned from either flank; however, these are rare. Desert terrain facilitates bypassing enemy positions and obstacles, but detailed reconnaissance must be performed first to determine if bypassing is feasible and will provide an advantage to friendly forces.

Dismounted infantry may be used to clear passes and defiles to eliminate enemy ATGM positions prior to the mounted elements moving through.

Avenues of approach of large forces may be constrained due to limited cross-country capability of supply vehicles coupled with longer lines of communications. The limited hard-surface routes that do exist are necessary for resupply.

RECONNAISSANCE

Reconnaissance is especially important in desert environments. Reconnaissance is a mission undertaken to obtain information by visual observation, or other detection methods, about the activities and resources of an enemy, or about the meteorologic, hydrographic, or geographic characteristics of a particular area. The desert environment may influence any or all of these techniques. The environmental effects on troops and their equipment may also influence observation techniques, or the frequency of vehicle and equipment maintenance that is required. Reconnaissance produces combat information. Combat information is a by-product of all operations, acquired as they are in progress. Reconnaissance, however, is a focused collection effort. It is performed prior to or in advance of other combat operations, as well as during that operation, to provide information used by the commander to confirm or modify his concept. Cavalry is the Army corps or division commander's principal reconnaissance organization.

Surveillance is a primary task of Army cavalry during reconnaissance operations. Surveillance is the systematic observation of airspace or surface areas by visual, aural, electronic. photographic, or other means. Scouts, ground and air, are the principal collectors of information. Scouts and their associated equipment are particularly affected by the environmental aspects of deserts. They require equipment that enhances their senses allowing them to conduct mounted and dismounted surveillance with stealth, at long-range, and in limited visibility, all of which can be adversely influenced by the desert environment.

SECURITY

Security operations obtain information about the enemy and provide reaction time, maneuver space, and protection to the main body. Security operations are characterized by aggressive reconnaissance to reduce terrain and enemy unknowns, gaining and maintaining contact with the enemy to ensure continuous information, and providing early and accurate reporting of information to the protected force. Security operations may be affected by various aspects of the desert environment including the sun, wind, sand, vegetation, sandstorms, terrain, and heat. Security operations include-

- Screen.
- Guard.
- Cover.

Counterreconnaissance is an inherent task in all security operations. Counterreconnaissance is the sum of actions taken at all echelons to counter enemy reconnaissance and surveillance efforts through the depth of the area of operation. It is active and passive and includes combat action to destroy or repel enemy reconnaissance elements. It also denies the enemy information about friendly units.

COMMAND, CONTROL, AND COMMUNICATIONS

The following paragraphs describe command, control, and communications considerations when operating in a desert environment.

Command

The effort to synchronize battlefield operating systems during the planning process can be negated by the failure to continue the synchronization effort during the preparation phase of a mission. This is especially true in the construction of engagement areas for defensive operations. Direct fire, indirect fire, and obstacles are linked, and the adjustment of one requires the adjustment of all. The commander must know and have a feel for what his unit can do, how long his unit takes to accomplish a mission, and what he really wants his unit to accomplish.

Adjustment of the elements of the battlefield operating systems can unravel the focus of a commander's intent. This is especially true in open terrain. Tactical commanders should personally direct the synchronization of engagement areas. Obstacles should be positioned, indirect fires adjusted, and direct fires rehearsed under the personal supervision of the commander. The commander controls operations using a highly mobile command group located well forward. He personally directs the battle, but must not be drawn into personally commanding an isolated segment of the force to the detriment of the remainder of the command. As previously mentioned, dry desert conditions can sometimes reduce radio signal strength and create unforeseen blind spots, even in aircraft operating nap of the earth.

Units may employ either a jump TOC or retransmission stations to facilitate communications with rear areas, as maneuver units are unlikely to be in one place very long. (If wire is used it should be buried to a minimum depth of 12 inches to avoid damage from track vehicles or shell fire.) There must be plenty of slack in the line to allow for sand shift and accurate map plots of buried wire should be kept. If overhead wire must be used, it should be mounted on posts erected in the form of tripods to avoid falling during severe weather.

Air or vehicle mounted liaison officers can be used if units are stationary or under listening silence. They should be proficient in navigation and sufficiently equipped to facilitate parallel planning. Liaison officers are highly effective and should be employed at every opportunity.

Continuous Operations

Continuous operations are affected by a number of factors in a desert environment. Fatigue is probably the foremost degrader of performance. Performance and efficiency begin to deteriorate after 14 to 18 hours of continuous work and reach a low point after 22 to 24 hours. Most tasks involving perceptual skills begin to show a performance degradation after 36 to 48 hours without sleep. Soldiers/marines cease to be effective after 72 hours without sleep. Performance decreases dramatically in an NBC environment and sleep becomes more difficult in MOPP gear. Sleep deprivation coupled with the environmental factors of the desert and the stresses of combat can significantly affect mission accomplishment.

The two categories of personnel who can be expected to show signs of fatigue first are young immature soldiers/marines who are not sure of themselves and seasoned old soldiers/marines upon whom others have relied and who have sustained them at cost to themselves. Commanders and leaders often regard themselves as being the least vulnerable to fatigue. Tasks requiring quick reaction, complex reasoning, and detailed planning make leaders the most vulnerable to sleep deprivation. Leaders denying themselves sleep as an example of self-control is extremely counterproductive. These factors are complicated by the environmental aspects of desert operations and should be considerations for operational planning.

Control

Clear identification of engagement areas is necessary to facilitate the massing and distribution of fires. In the absence of identifiable terrain, target reference points (TRPs) can be created with damaged/destroyed vehicles that are moved into required locations at the direction of commanders invested with the responsibility for specific engagement areas. Other types of TRPs could be used. For example, marker panels, visible and infrared chemical lights, flags, and white phosphorus/illumination rounds could be used. The construction or fabrication of TRPs must be resourced and well planned in order to be effective. For example, how will TRPs be replaced for subsequent defensive operations? Another common problem is TRP proliferation, which makes TRPs difficult to identify when each echelon of command has allocated too many TRPs.

Pyrotechnics are usually more effective in desert climates than in temperate climates; however, heat mirages and duststorms may impair or restrict their use. Even heliographs (signal mirrors) may be useful as they are directional and therefore can aid security. Sound communications are usually impractical due to distance, vehicular noise, and storms, but can be used for local alarm systems.

Colored flags with prearranged meanings can be used as a means of communication in the flat open terrain of the desert. Colored flags tied to antennas may also assist in vehicle/unit recognition during limited visibility operations and offensive operations.

As previously described, the desert offers excellent fields of fire. Tanks and heavy antitank weapons should be sited to take advantage of their long range and accuracy. Firing first and accurately are the most important considerations in desert operations.

Target identification is the recognition of a potential military target as being a particular object (such as a specific vehicle, by type). At a minimum, this identification must determine the potential target as friendly or threat (identify friend, foe, or neutral [IFFN]). Because it is easy to become disoriented, it is often necessary to mark sectors of fire on the ground with poles or rocks, if available.

Communications

Communications support is also adversely affected by high temperatures. The heat causes anomalies in radio and other electrical transmissions, and radio battery life is reduced. Radio range is shorter during the day than at night. At night, range improves but static electricity may cause interference. FM communications range can be reduced by as much as 50 percent because of high temperatures. HF ground wave propagation over the dry sandy soil is reduced.

Night communications make communications security a concern, as it always should be. Experience in Desert Shield and Desert Storm indicates vastly expanded ranges of FM radios. Communications between units 40 to 50 kilometers apart was not unusual. Communications obviously affect command and control as well as intelligence collection and dissemination, and their importance must not be underestimated.

COMBAT SUPPORT

A force operating in the desert must be a balanced force with combat support and combat service support-it must be a combined arms team. While principles of combat support operations are found in doctrinal manuals dealing with a specific arm of service, there are some techniques that must be modified or emphasized in the desert.

INTELLIGENCE

The relative importance of intelligence sources may vary from that expected in more conventional areas. Enemy prisoners of war require immediate interrogation as the flexibility of operations will rapidly make their information outdated. Information given by civilians encountered in desert operations should be treated with caution unless corroborated. Military intelligence teams located in the area of operations can determine if these EPWs and civilians are in fact what they say they are, or infiltrators sent to harass the rear area and commit acts of sabotage. Electronic support measures are a major source of intelligence in desert warfare. Enemy activity, or the lack of it, is a good source of information; so punctual, accurate reports by all sources, both positive and negative, are necessary.

FIRE SUPPORT

The Allies in North Africa in 1942 found that placing small field artillery units in support of small maneuver units gave the units a sense of security, but produced limited results. Field artillery was effective only when massed (battalion or higher) and only when continued for some time because of the protective posture and mobility of the target. Typically, the control of massed fires was the responsibility of the division artillery.

The Allies in North Africa in 1942 experienced heavy casualties from Axis units overrunning the artillery positions after penetrating the armor and infantry positions. Often, the Axis units would attack from the east at one time, from the west later, and from several directions simultaneously. At first, the Allies simply emphasized direct fire. Later, the Allies attached antitank gun units to the artillery battalions to increase the artillery's antitank ability.

When armor and infantry units move, the artillery must move with them. The most useful technique is for the artillery to move in a formation with a lead vehicle so that, immediately upon stopping, the artillery is in a position or formation to deliver fire in any direction and simultaneously defend the position from any direction. The Allies in North Africa in 1942 and units in Desert Shield/Storm found that the armor and infantry units would outdistance the artillery unless the artillery moved with them. The artillery moved within 2-3 kilometers of the leading troops to provide responsive fires. The armor and infantry provided protection for the artillery. The whole group moved in one cohesive formation, sometimes in a large box or diamond formation.

Due to the fluid nature of desert operations and the possibilities for excellent enemy observation, close and continuous field-artillery support for all levels of the force is necessary. Field-artillery pieces should be at least as mobile as the force they are supporting. Crews must be proficient in direct fire and prepared to defend against a ground attack.

Due to the threat of immediate counterbattery fire, field artillery units must be prepared to move into position, fire, and rapidly displace to another position. A battery should be prepared to displace several times a day.

Field artillery units employed in desert operations should be equipped with the most sophisticated survey devices available. Manual systems are slower and not necessarily as accurate, thus affecting tactical employment and reducing response time.

Aerial observation may often be extremely difficult due to enemy air defense, so most adjustment is by ground observers. How the environment affects observation of fires was described previously in this chapter in the paragraph, "Observation and Fields of Fire." Recompute weather conditions frequently as weather conditions can change rapidly from the morning to the evening, and thus affect the accuracy of fires.

Fires are planned as in temperate climates. When there are no significant terrain features along a route of advance, targets are planned using coordinates.

A moving force in a desert is at a disadvantage in comparison with a stationary unit due to lack of concealment and the presence of dust clouds. The defender may engage with missiles from an unexpected direction or from terrain features of no apparent significance. The attacker must be prepared to rapidly shift fires to suppress unforeseen targets. Tactical aircraft may be used to suppress or destroy targets. Targets for aircraft can be marked with indirect- or direct-fire smoke. White phosphorus or illuminating rounds set for low-air burst are also effective.

Indirect fires are used to slow the enemy advance, to suppress enemy weapons and observers, and to conceal movement between positions using smoke. Defensive operations in deserts are characterized by long-range engagement with tanks and ATGMs.

AIR DEFENSE

Identification of friend or foe is difficult. Throughout the entire theater of operations there will be numerous weapon systems that are common to both sides of the conflict. The individual soldier/marine is going to be faced with the monumental task of separating friend from foe by more than just from the recognition of the manufacturer or silhouette of a piece of equipment. This will be true of both air and ground systems. This identification problem will be compounded by the nonlinear battlefield where the focus of operations will not be separated by a line.

The desert is an outstanding environment for employing aircraft. Every unit must be extremely proficient at passive and active air defense. The Allies in North Africa and the Israelis in the Middle East found that dispersion limited the effects of air attacks, and small arms air-defense techniques were effective. Almost every weapon in North Africa had a secondary antiaircraft or antitank mission.

Emphasize to each unit that, when in position, units must disperse very widely making a less lucrative target. When moving in column and under air attack, units must move at least 40 to 50 meters off the road because aircraft normally have nose guns trained on troth sides of the road. A vehicle on the road or on both sides of the road will die.

Because of the wide open spaces characteristic of many deserts and the relatively large areas associated with desert operations, forces fighting in the desert should be reinforced with additional air-defense weapons. Still, there may not be sufficient dedicated air-defense systems to fully cover the force. When this is the case, commanders must be especially careful when establishing air-defense priorities in view of relatively long lines of communication and the tendency to maneuver over relatively large areas. In any event, all units must include a scheme for countering air attacks in their battle plans using both active and passive measures.

Although Army armored and mechanized infantry division air-defense weapons are tracked, this does not necessarily apply to corps medium-altitude air defense units. However, Army corps surface-to-air missile (SAM) units have considerably greater ranges and are equipped with more sophisticated early warning and control systems. Some corps units should be employed well forward. These weapons will have to displace by section to ensure continuous coverage.

Air-defense units should be located close to elements of supported units to provide for ground defense. When the supported unit moves, the air defense unit must also move, which requires careful coordination to ensure that movement of the supported unit is not delayed. Airspace management difficulties are compounded in the multinational environment. SOPS should be exchanged among multinational forces to lessen the confusion of airspace management.

ENGINEERS

Engineer operations in the desert are similar to those in temperate climates although there are fewer natural terrain obstacles to be crossed. Depending on the terrain anticipated in the operations area, a dry-gap crossing capability may have to be obtained from corps support units. Important tasks for engineers in desert operations include-

- Mobility/countermobility/survivability support, including construction of obstacles, logistics facilities and routes, field fortifications, airfields, and helicopter landing pads.
- Water supply.
- Topographic support (map-making).

Mobility

The vastness of the desert makes mobility a prime concern. Roads are usually scarce and primitive. Cross-country mobility is possible in some areas, but trafficability is poor in soft sand, rocky areas, and salt flats. Engineers assist maneuver by reducing slopes, smoothing rock steps, and bridging dry gaps.

Expanded engineer reconnaissance capability will be needed to identify routes, existing obstacles, and minefield locations. Flat, open areas provide good sites for aircraft landing strips; however, in most cases the soil must be stabilized. Normally, desert soil produces extensive dust and has limited weight-bearing capacity.

Engineers use various agents to alleviate severe dust conditions (diesel, JP4, or oil mixtures for example). This is particularly critical in reducing engine wear in areas supporting rotary wing aircraft. It is also important along heavily traveled roads and in cantonment areas. Engineers also use soil-stabilization techniques to increase soil-bearing capacity for airstrips and MSRs.

The application of the fundamentals of breaching-suppress, obscure, secure, and reduce-and the organization of the force in terms of supporting, breaching, and assaulting elements, are even more important in the desert due to the enhanced observation and fields of fire. However, the desert does offer greater opportunities to bypass enemy obstacles because of the greater range of mobility afforded by desert terrain. Caution must be exercised when choosing to bypass enemy obstacles since the bypass may lead the force to the enemy's kill sack.

The increased mobility in the desert makes it easier for the enemy to counterattack exposed flanks of attacking forces. Plan obstacles to protect flanks during offensive operations. Beyond conventionally emplaced minefield, FASCAM, which includes artillery-delivered mines, GEMSS, and air-delivered Gator munitions, are all systems that lend themselves to situational development. FASCAM and conventional minefield may be appropriate, but consider the time required to employ FASCAM when selecting this option. Artillery-delivered FASCAM does not deploy well in soft sand and removes a majority of your indirect-fire assets from the fight.

Countermobility

Due to the mobility inherent in desert operations, obstacles must be extensive and used in conjunction with each other and with any natural obstacles, and covered by direct and indirect fires. Isolated obstacles are easily bypassed.

Mines are easily emplaced in a sand desert, and blowing sand will effectively conceal evidence of emplacement. However, the following potential problem areas must be considered when emplacing mines:

- Large quantities of mines are required for effectiveness.
- Sand can cause malfunctioning.
- Shifting sand can cause mine drift.
- An excessive accumulation of sand over the mines can degrade performance.
- Sand may be blown away and expose the mines.

In suitable terrain, antitank ditches that exceed the vertical step of enemy main battle tanks may be used. Because antitank ditches cannot be conceded, they must be dug so they do not outline a defensive front or flank. They have the advantage of not requiring as much logistic support as minefield. They must be covered by observation and fire to prohibit enemy infantry using them as ready-made trenches.

Because of limited off-road mobility of most combat service support vehicles, considerable engineer efforts may be necessary to construct and maintain routes forward to maneuver units. Local resources, such as salt-marsh mud laid on sand can be used. Track vehicles should not use these routes since they could easily ruin them.

Most desert regions have a natural terrain structure that restricts maneuver such as sandy dunes, rocky plateaus, mountains, and wadis. These structures must be interpreted rapidly and correctly, and then reinforced with obstacles to fix, turn, or disrupt enemy movement, according to the commander's plan.

Minefield and antitank ditches are the primary means of creating obstacles in the desert. Antitank ditches require extensive preparations, but they are effective when adequate preparation time is available. Many desert villages have irrigation ditches that can be used tactically. Other countermobility methods are generally not effective. Road craters, for example, are usually easy to bypass. In sandy areas, ditches can easily be filled in, so they are not good obstacles. Opportunities for bridge destruction are rare, and local materials for expedient obstacles are scarce.

Engineers and combat forces should coordinate the siting of planned obstacles to support the defensive concept. In defensive operations the effectiveness of obstacles requires synchronization.

Survivability

Desert terrain varies from region to region. Generally, however, observation is excellent and concealment is difficult. Deserts provide little cover and concealment from ground-based observers and even less from aircraft. These conditions make modem weapon systems more lethal in deserts than in any other environment.

In the desert, hull and turret defilades for tactical vehicles are essential. This allows the defending force to take advantage of their long-range weapon systems in the face of enemy fires. Dispersion and frequent moves are other survivability techniques that can be used.

The preparation of fortifications in the desert is difficult. Fortifications in sandy soil often require revetments. In rocky plains or plateaus it may be impossible to dig. To counter this problem, build up emplacements with rocks and use depressions.

Camouflage is very effective when properly employed; however, patterns and techniques must be carefully selected to match the local desert environment. Camouflage nets should be provided for all equipment. See Appendix E for additional comments on desert concealment and camouflage.

Desert Survivability Positions

Defensive positions are very vulnerable to offensive fire due to long-range observation and fields of fire in the desert. This, coupled with a lack of natural obstacles, may lead the commander to invest the bulk of his engineer effort into survivability positions. Survivability positions enhance the ability of all direct-fire elements to survive indirect-fire and to return fire on the enemy. Survivability positions are normally more important than antitank ditches, especially in open terrain. See Figures 3-1 through 3-6, for examples of survivability positions. The following are some things you should or should not do when preparing survivability positions:

DO-

- Ensure adequate material is available.
- Dig down as much as possible.
- Maintain, repair, and improve positions continuously.
- Inspect and test position safety daily, after heavy rain, and after receiving direct and indirect fires.
- Revet excavations in sandy soil.
- Interlock sandbags for double-walled constructions and corners.

DO NOT-

- Fail to supervise.
- Use sand for structural support.
- Forget to camouflage.
- Drive vehicles within 6 feet of a position.
- Overfill sandbags.
- Put troops in marginally safe bunkers.
- Take shortcuts.
- Build above ground unless absolutely necessary.
- Forget lateral bracing on stringers.

- Check stabilization of wall bases.
- Fill sandbags approximately 75 percent.
- Construct to standard.
- Use common sense.

The commander's responsibilities during construction of survivability positions are to—

- Protect troops.
- Continuously improve and maintain unit survivability.
- Provide materials.
- Periodically inspect.
- Plan and select fighting position sites.
- Get technical advice from engineers, as required.

In a combat situation, it may be necessary to improvise construction of a survivability position by using materials not normally associated with the construction. Some examples of field-expedient materiel are-

Wall Revetment

- Sheet metal.
- Corrugated sheet metal.
- Plastic sheeting.
- Plywood.
- Air mat panels.
- Air Force air-load pallets.

Overhead Cover Stringers

- Single pickets.
- Double pickets.
- Railroad rails.
- "I" beams.
- Two-inch diameter pipe or larger.
- Timbers 2" x 4", 4 " x 4 ", and larger.
- Reinforced concrete beams.
- 55-gallon drums cut in half longitudinally.
- Large diameter pipe/culvert, cut in half.
- Precast concrete panels, 6-8 inches thick.
- Airfield panels.
- Air Force air-load pallets.

Wall Construction (Building up)

- Sand-grid material.
- 55-gallon drums filled with sand.
- Expended artillery shells filled with sand.
- Shipping boxes/packing material.
- Prefabricated concrete panels.
- Prefabricated concrete traffic barriers.

Stand-Alone Positions

- Prefab concrete catch basins, valve pits, and utility boxes.
- Military vans.
- Connexes or shipping containers.
- Large diameter pipe/culvert.
- Steel water tanks.
- Other storage tanks (cleaned and ventilated).
- Vehicle hulks.

- Shipping pallets.

The following is a suggested inspection checklist to follow when preparing survivability positions:

- Location is sited tactically sound.
- Low profile is maintained.
- Materials are of structural quality (standard construction material).
- Excavation-walls are sloped.
- The setback for overhead is a minimum of 1 foot or 1/4 the depth of cut.
- Stringers¬
  - Are firmly on a structural support.
  - Have lateral bracing emplaced along supports.
- 2" x 4" or 2" x 6" stringers are used on the edge; the strength is on the depth of the lumber.
- Supports¬
  - Stringers are firmly on supports.
  - Supports extend past the excavation by 1/2 the depth of cut.
- Revetments—
  - Quality of construction is checked.
  - Sheeting is supported by pickets.
  - Pickets are tied back.
- Overhead cover¬
  - Quality of structural layer is inspected.
  - Quality of dust layer-plywood or panels-is inspected.
  - Layer is cushioned at least 18 inches deep.

The one-man fighting position is the individual's basic defensive position. The one-man fighting position with overhead cover (see Figure 3-1) provides protection from airburst weapon fragments. A good position has overhead cover that allows the soldier/marine to fire from beneath it. Stringers extend at least 1 foot on each side of the position to provide a good load-bearing surface for overhead cover.

Figure 3-1. One-man Fighting position with overhead cover.

Generally, the two-man fighting position is preferred over a one-man position since one soldier/marine can provide security while the other is digging or resting. The position can be effectively manned for longer periods of time; if one soldier/marine becomes a casualty, the position is still occupied. Further, the psychological effect of two men working together permits occupation of the position for longer periods. Overhead cover also improves the position's effectiveness; it is made as described for the one-man position (see Figure 3-2).

Figure 3-2. Two-man fighting position wiih overhead cover.

Fighting positions for machine guns are constructed so the fires are to the front or oblique; the primary sector of fire is usually oblique so the gun can fire across the unit's front. The position is shaped so the gunner and assistant gunner can get to the gun and fire it to either side of the frontal direction. Overhead cover is built over the middle of the position (see Figure 3-3). It is constructed as described for the one-man position.

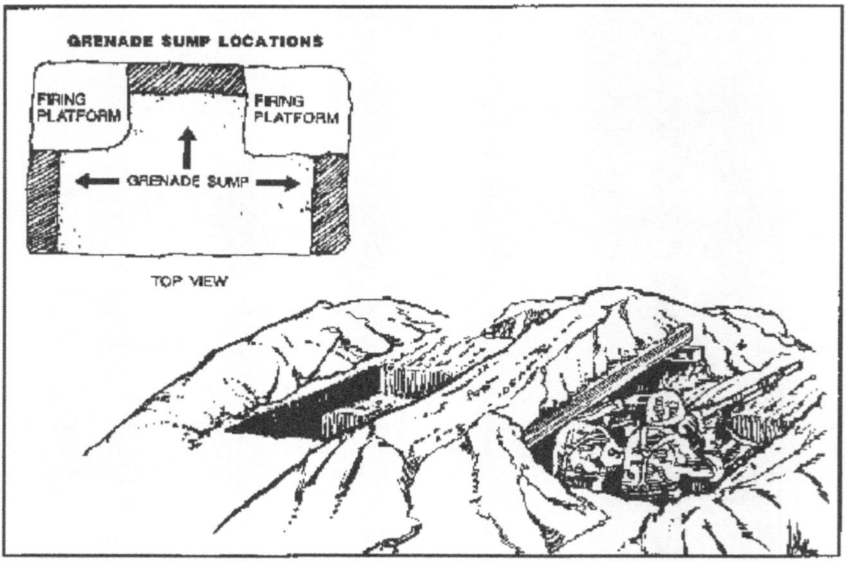

Figure 5-3. Machine-gun position with overhead cover.

Protective shelters and fighting bunkers are usually constructed using a combination of the components of positions mentioned thus far. Protective shelters are primarily used as command posts, observation posts, medical aid stations, supply and ammunition shelters, and sleeping or resting shelters. Figure 3-4 shows an example of a command bunker.

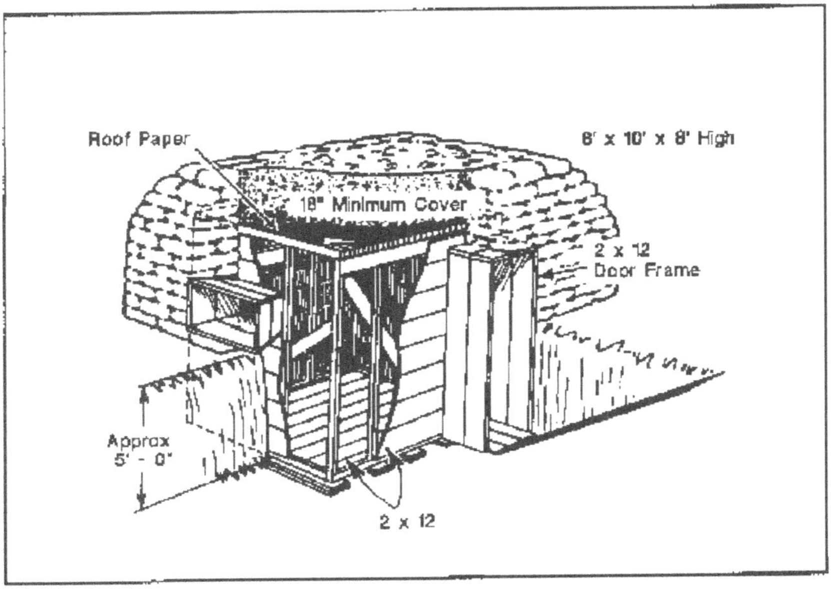

Figure 3-4. Command Bunker .

The Dragon position requires some unique considerations. The soldier/marine must consider the Dragon's extensive backblast and muzzle blast, as well as cleared fields of fire. When a Dragon is fired, the muzzle extends 6 inches beyond the front of the position, and the rear of the launcher extends out over the rear of the position. As the missile leaves the launcher, stabilizing fins unfold. Therefore, the soldier keeps the weapon at least 6 inches above the ground when firing to leave room for the fins. A waist-deep position will allow the gunner to move while tracking a target. Because of the Dragon's above ground height, soldiers/marines should construct frontal cover high enough to hide the soldier's/marine's head and, if possible, the Dragon's backblast. The soldier/marine must dig a hole in the front of the position for the biped legs. If cover is built on the flanks of a Dragon position, it must cover the tracker, missiles, and the gunner. Overhead cover that would allow firing from beneath it is usually built if the backblast area is clear (see Figure 3-5).

Figure 3-5, Dragon Position.

A fighting position for the dismounted TOW must not interfere with the launch or tracking operations of the weapon. As with Dragon and LAW positions allowances for backblast effects are necessary. Backblast and deflection requirements restrict the size of overhead cover for the weapon. See Figure 3-6.

Figure 3-6.Dismounted TOW position.

Designers of fighting positions and protective positions in desert areas must consider the lack of available cover and concealment. Fighting positions should have the lowest profile possible, but mountain and plateau deserts have rocky soil or "surface chalk" soil which makes digging difficult. In these areas, rocks and boulders are used for cover. Because target acquisition and observation are relatively easy in desert terrain, camouflage and concealment, as well as light and noise discipline, are important considerations during position construction.

Indigenous materials are usually used in desert position construction. However, prefabricated structures and revetments, if available, are ideal for excavations. Metal culvert revetments can be quickly emplaced in easily excavated sand. Sandbags and sand-filled ammunition boxes are also used to prevent side walk of positions from collapsing.

FM 5-103 discusses vehicle fighting positions in detail and should be consulted for more information. Figure 3-7 provides specifications for vehicle survivability defilade positions that can be dug by the D-7 dozer.

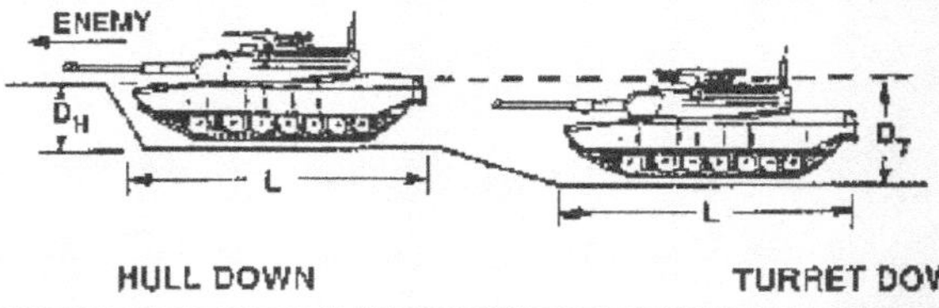

VEHICLE	L	D_H	D_T	WIDTH
M1 Tank	2 Dozer lengths	Top of battery box	Top of your head	1 1/2Blade widths
M60 Tank	2 Dozer lengths	One fist above battery box	Top of smoke stack	1 1/2 Blade widths
M2 BFV M3 CFV	1 1/2 Dozer lengths	Top of engine housing	Top of smoke stack	1 1/2 Blade widths
M113 APC	1 1/2 Dozer lengths	One fist above battery box	Top of air intake cover	1 1/4 Blade widths
M901 ITV /FISTV	1 1/2 Dozer lengths	Top of engine housing	Top of your head	1 1/4 Blade widths
M109 155 Howitzer	6 1/2 Dozer lengths	One fist above battery box	Not required	1 1/2 Blade widths
M110 8" Howitzer	6 Dozer lengths	Top of blade tip crown	Not required	1 1/2 Blade widths
M577 Command Post Vehicle	1 1/2 Dozer lengths	Top of your head	Not required	1 1/4 Blade widths

These dimensions are described with respect to you and your dozer. Techniques for construction of survivability positions are described in FM5-34, ENGINEER FIELD DATA.

Figure 3-7. Survivability defilade positions.

Logistics areas (BSA/DSA) require additional survivability support. Desert operations require that logistics concentrations such as BSAs and DSAs be given additional considerations for survivability support. These sites are large, relatively static, and difficult to camouflage. As a result, these support areas are vulnerable to enemy interdiction. Military vans or connexes should be covered with sandbags to improve protection. Additionally, if they are covered with heavy plastic, with plastic drapes over the entrances, protection against NBC effects can be improved.

Engineer digging assets, such as bulldozers, should be tasked to provide survivability support to these sites with particular emphasis placed on hardening ammunition and fuel storage locations. Caution should be used when digging foxholes and tank hide positions since some areas have a tendency to cave in.

Water Supply

Water supply is the most important mission of engineers in the desert. The search for water sources requires continuous, intensive reconnaissance. Water may be obtained by drilling beds of dry water courses, or by deepening dry wells. Once found, water must be made potable and stored or transported. Since water purification trucks may be high-priority targets and barely sufficient for the task, any force operating in the desert must be augmented with water supply units (including well drilling), water purification and water distillation teams, and transportation teams. Another possible water source is the reverse osmosis water purification unit (ROWPU). This unit is an ISO frame-mounted, portable water purification system capable of purifying water from almost any shallow well, deep well, and surface water or raw water source. The ROWPU is capable of removing NBC contaminants, minerals, and biological impurities. The single greatest benefit of the reverse osmosis process is the ability to desalinate sea water. The ROWPU is capable of producing potable water at a rate of 600 gph. The ROWPU is powered by a 30-kilowatt generator set.

Topographic Support (Map-making)

Large areas of the world's deserts are not covered by maps of any useful tactical scale. Existing maps are frequently inaccurate and increase the difficulties of navigation. Therefore, engineer topographic companies must augment the force by preparing, printing, and distributing up-to-date maps of the operational area. USAF, Army, and Marine aviation support can be used to produce gridded maps from aerial photography of the area forward of the line of contact.

MILITARY POLICE

Combat support provided well forward by military police will continue in desert operations, although over increasingly extended distances. MP tactical and physical security will be of special importance over extended lines of communication, such as petroleum pipelines and viaducts transporting water over long distances. Protection of these critical items demands both active and passive measures, including overflight by returning aircraft or overwatch by convoy movements. The storage sites for water, food, POL, and ammunition have historically been principal targets for enemy action, and consequently must receive augmented security.

The indefinite conditions and number of roadways will require increased circulation control points to direct traffic, redirect stragglers, and provide information so that throughput forward to the fighting elements will be expedited. Military police are especially valuable when the combat commander must employ concentration or economy of force in the face of the enemy to gain a favorable combat ratio. MPs can secure the roadways, enforce priority movement, and prevent any delay of the elements undertaking passage of lines to blocking or defensive positions. MPs can also assist in the handling of EPWs.

US AIR FORCE SUPPORT

US Army and Marine forces fighting in the desert can expect to be supported by USAF tactical fighter-bomber and airlift aircraft. Close air support by USAF tactical fighter bombers is more important in desert warfare in view of lack of concealment, relatively large areas of operations, and mobility of forces employed by each side. Air support in a desert environment has advantages over more temperate areas of operations. For example, it is easier to locate targets; visual observation is normally far superior to that in temperate climates; and ground movement is more readily apparent.

Air attacks can be handicapped by lack of covered approaches, but increased visibility permits engagement from standoff ranges. When flying close air support missions it is important for pilots to be able to differentiate between enemy and friendly forces. Use panels or other visual or electronic identification means to assist in identification.

Because of the extended lines of communication likely in desert operations, USAF tactical airlift should be used whenever possible. This is particularly true of resupply operations conducted from a lodgement area to forward trains areas when considerable distances are involved.

Planning for air support must be as detailed as time permits to determine mission and armament requirements, time over target, and method of control. The joint air-ground operations system (AGOS) used to request and coordinate the use of US Air Force tactical air support is described in FM 100-26.

US NAVY SUPPORT

When the force is being supported by US Navy gunfire, or Navy or Marine aircraft, elements of a Marine air and naval gunfire liaison company (ANGLICO) are attached to Army ground forces. The mission of the company is to support an Army division by providing control and liaison agencies for employment of this support.

ANGLICO platoons and teams can advise commanders on the capabilities, limitations, and employment of naval gunfire, and USN or USMC air support. Platoons are normally placed with brigades or higher headquarters, and air and gunfire support teams placed with battalion task forces. Although the company has organic vehicles and some combat services support capability, its elements generally require additional administrative and logistical assistance from the supported unit. In order to communicate with Army/Marine units, additional communications equipment may also have to be provided. Additional information on ANGLICO employment can be found in FM 31-12.

COMBAT SERVICE SUPPORT

Combat service support for desert operations is described in detail in Chapter 4. When planning a desert operation it is necessary to consider the following factors:

- The speed of supply may be slowed and lines of communication can be vulnerable due to the distances between units. Except for Class V and sometimes Class III, resupply should beat night for reasons of security.
- A great demand for water can tie down large quantities of transport and may involve laying pipelines. Water is vital, so consider the water situation during every operation estimate.
- Increased maintenance is required due to heat, sand, and dust damage to equipment. This not only increases the repair work load, but also increases demand for replacement items due to increased wear.

Supply

Offensive operations in this environment may involve considerable expenditure of ammunition and high POL consumption. Units must carry maximum combat supplies, and plans for resupply must be widely disseminated and clearly understood. Use every opportunity for resupply.

Due to the importance of combat service support, attacking the rear will be more immediate y effective at lower levels of command in the desert than in temperate climates. In the case of water, for example, the enemy must be able to obtain resupply. Degradation of the enemy's trains places him in a situation where his troops must maneuver against the attacker regardless of the planned scheme of defense.

Combat service support must be reliable and timely, using vehicles that can travel over difficult terrain to reach combat units. In the desert, more than anywhere else, the commander must ensure that he has support that is capable of maintaining his unit for a specified period of time, even if the logistic line of communication is temporarily broken.

The mobility and freedom of tactical maneuver are tied to the ability of the logistic chain to supply maneuver units. Two alternatives are available: increase the rate of supply, probably requiring more vehicles, or prestock, which ties units to the stocked area. Some important supply considerations are outlined below:

- Class I. It is often impractical to supply hot rations from mess trucks, especially when the unit is subject to enemy air reconnaissance or target acquisition devices. T rations and B rations are the usual method of troop feeding.
- Class III. Daily requirements for POL in desert operations can be expected to be high. Estimates for POL requirements should take into consideration large-scale maneuver inherent in desert operations.
- Class V. Estimates of ammunition requirements should reflect the high level of commitment that can be anticipated in desert operations.

Maintenance

Disabled vehicles are vulnerable targets. Both disabled vehicles and the maintenance vehicles used in working on them must be concealed during the day, and strict light and sound discipline imposed at night. Maintenance contact teams should carry Class IX supplies that have a quick turnover.

Section II. Offensive Operations

This section discusses offensive operations as they are modified by desert terrain.

GENERAL

The main purpose of offensive operations in desert terrain is to destroy the enemy, Operations may be undertaken to secure key or decisive terrain, to deprive the enemy of resources or decisive terrain, to deceive and divert the enemy, to develop intelligence, and to hold the enemy in position. Destruction of the enemy can be accomplished by concentrating friendly forces at a weak point in the enemy's defense and destroying enemy combat units, or by driving deep into the enemy's rear to destroy his combat service support and cut his lines of communication. No force can survive in the desert for long without combat service support.

An imaginative commander is not bound by terrain constraints in seeking and destroying the enemy. Due to the scarcity of key terrain in the desert, normally the only constraints placed upon a maneuvering force is its ability to maintain the only constraints placed upon a maneuvering force is its ability to maintain responsive combat service support and to protect its combat service support from enemy attack. The longer the lines of communication become, the more susceptible they are to being cut.

In most deserts, the scarcity of large areas of defensible terrain means that an enemy force has at least one flank open to attack, The attacking force must seek this flank and attempt to maneuver around it into the enemy's rear before the enemy can react and block the envelopment with mobile reserves.

Successful offensive operations depend on rapid, responsive, and violent maneuver, seeking a vulnerable enemy flank while exposing none to the enemy. The enemy, realizing the danger of remaining stationary in this terrain, may choose to conduct spoiling attacks or to counterattack. The resulting meeting engagement between the two attacking forces will often be a series of flanking actions and reactions with success going to the one who can find the other's unguarded flank first.

Attacking forces may conduct or participate in movement to contacts or hasty or deliberate attacks. Within a division, lead elements of forward units may be conducting a deliberate attack on the enemy's weak point or flank to open a gap for following units to move through and exploit success. Lead units of the exploiting force will be conducting a movement to contact and hasty attacks to overcome pockets of enemy resistance. Regardless of the type of operation being conducted, attacking units use the fundamentals for offensive operations described in FM 100-5/FMFM 6-1 and other tactical manuals, modified to suit the terrain.

FUNDAMENTALS OF THE OFFENSE

The attacker must conduct active and aggressive reconnaissance to the front, flanks, and rear, not only to locate and identify enemy obstacles, units, weak points, and flanks, but also to give early warning of threats to his flanks and combat service support elements. A moving force is at a disadvantage in the desert due to a lack of concealment. Therefore, it is necessary to push reconnaissance units as far out from the main body as possible to allow early warning and to deny the enemy close-in observation.

Information gathered by this reconnaissance must be passed promptly to all units. In the desert, a negative report may be as important as an enemy sighting. Commanders and staffs must avoid the two extremes of either passing too little information or overwhelming their subordinates with useless trivia. Similarly, reconnaissance units must also avoid extremes. There is a very real possibility that extensive reconnaissance in one area will alert the enemy of intended operations in that area. Therefore, the need for reconnaissance must be tempered with the need for deception. In fact, reconnaissance may even serve as a deceptive measure to draw the enemy's attention away from the real objective or area of operations.

Concentrate on overwhelming combat power. Mass is achieved in both time and space. Units must be able to rapidly concentrate at a given time and place, and then disperse just as rapidly to avoid offering a lucrative target to the enemy. Concentration does not necessarily mean that vehicles and men are massed in a small area, but that units have the ability to place an overwhelming concentration of fires on the enemy.

Mutual support is as important in the desert as in temperate climates. Due to the large distances covered by maneuver in the desert, mutual support does not mean that any one unit is always in position to fire against an enemy threatening another unit. However, units must be capable of maneuvering in support of one another without disrupting the scheme of maneuver.

Concentration requires movement, and possibly weakening of forces facing the enemy in another part of the zone. Due to the enemy's observation capabilities, movement should take place at night or in conditions of limited visibility whenever possible. Deception measures play an important part in concentration, either to mislead the enemy as to the strength or true intentions of the opposing forces, or their avenues of approach. In this environment of negligible concealment, deception cannot be overemphasized.

The enemy's objective is to stop and destroy the attacking force by direct and indirect fires, obstacles, and counterattacks. The attacker must in turn suppress enemy weapon and surveillance systems to degrade their effects and their intelligence-gathering capability.

Attack helicopters and high-performance aircraft are extremely useful due to their ability to maneuver and apply firepower over a large battlefield in a short time. So, suppression of enemy air defense has a high priority during offensive operations. The destruction of enemy antitank capabilities must also have a high priority due to the shock potential of armor in the desert. No target that has a long-range antitank capability should be disregarded. Good gunnery and well-planned fire distribution are preeminent.

In featureless desert terrain, the requirement to shock, overwhelm, and destroy the enemy demands accurate reconnaissance to identify actual positions from false positions, and excellent navigation so that a commander may be certain of the deployment of his forces. Reconnoiter to find a gap or assailable flank (without alerting the enemy that the area is being reconnoitered) and concentrate the main body to go through or around it with suppressive fires on the flank(s). A gap must be wide enough to allow one unit to bypass another unit that could be stalled. Obstacles are likely to be placed so that attempts to go around them will often lead the attacker into a tire sack. Equipment capable of breaching obstacles must be located well forward.

ENVIRONMENTAL CONSIDERATIONS

As a general rule, a force attacking in daylight should try to wait until the sun is comparatively low and position behind it. This enables enemy targets to be plainly seen without their shadows, while the defenders are handicapped by glare, mirages, and haze. It is not always possible (nor essential) for the sun to be directly behind the attackers. To rely on this leads to a stereotyped method of attack which could become evident to the defenders. The commander of a maneuver force should attempt to keep the sun somewhere on a 3,200-mil arc to his flanks or rear, giving a wide choice of angle of attack.

Dust is an observational hazard to a maneuvering force, especially where there is little or no wind. Teams should move in echelon with overmatching elements on the upwind side, and observers and attack helicopters should operate well to the flank. Since it is impossible to disguise movement during daylight, the assault should be as rapid as possible to minimize enemy reaction time.

The decision to move through a sandstorm will depend on the unit's distance from the enemy, trafficability, the presence of minefield, and the direction and density of the storm. If the advancing unit is caught in a storm blowing from the enemy's direction, the safest alternative is to halt until it abates, although this may not always be possible. In some situations it may be possible for platoons to form close column, using taillights only, and continue movement. When the storm is blowing toward the enemy it is possible (and extremely effective) to conduct an attack immediately behind the storm.

In certain circumstances equipment or positions that are camouflaged and are less than 1 meter from the ground are invisible to an observer at the same height out to approximately 2,000 meters. At the same time, mirages allow observation of objects below the horizon, although these may be distorted, enlarged, or fuzzy to the point of being unrecognizable. These effects often depend entirely on the angle of the sun to the observer and are best combated by-

- Maintaining observers as high above the desert floor as possible, even if only in hull-down positions behind sand dunes.
- Allowing a vehicle's crew on one side of a position to warn a crew on the other side of a possible threat to his front by crews observing over wide areas.

Many offensive operations take place at night. Observation in these conditions varies according to the amount of ambient light. During nights when the moon is full or almost full, the clear desert sky and ample ambient light allow good observation, both with the naked eye and with night observation devices. Maneuvering units using night-vision devices must continually scan the surrounding terrain to pickup enemy activity that normally would be acquired by peripheral vision in the daylight.

The desert night is extremely dark when there is little or no moon. Under these conditions passive-vision devices, with the exception of thermal imagery, are of little value unless artificial light is used. Active light sources will have to be relied upon. Employment of artificial light must be strictly controlled by the headquarters directing the operation to maintain surprise. As a general rule, direct-fire weapons should not illuminate their target themselves, as their vision will be obscured by debris kicked up due to muzzle blast. Following contact, when some targets should be on fire, passive devices can be used.

MANEUVER

If the terrain permits masking of maneuvering units, and trafficability is good, normal fundamentals of fire and maneuver are used. Trafficability may be restricted by rocky terrain as in the Golan Heights, or the ground may be so flat that the defender has total observation of the area. Movement in these circumstances requires speed of maneuver, deception, and considerable suppression to degrade enemy observation and fires. Frontal attacks should be avoided, especially in conditions of restricted trafficability. It is preferable to maintain pressure on enemy units in unfavorable terrain, while other forces find enemy weaknesses in terrain that is more favorable for an attack.

Lack of clearly defined terrain features complicates navigation and phased operations. Units conducting an enveloping maneuver are apt to lose direction unless routes have been carefully reconnoitered by the maximum number of leaders.

Section 111. Defensive Operations

This section discusses defensive operations as they are modified by desert terrain.

GENERAL

It is unlikely that a US force will be fully deployed in a desert country before an enemy attack. The more probable situation, assuming a secure lodgement area, will be that part of the force will be in position supporting an allied force, while the remainder is moving in by air and sea. Tactically, the allied force will be outnumbered, so the initial mission will be to gain time until the entire force is present in the operational area. This will require a defensive posture initially, but a defense undertaken so aggressively as to convince the enemy that his offensive action will be too costly in personnel and equipment to be worth maintaining. The enemy will be well aware that US forces are arriving in the area, and will make every effort to conclude his operation successfully before the force is fully prepared for combat operations.

The force may conduct defensive operations during subsequent stages of the operation for any of the reasons described in FM 100-5/FMFM 6-1. Portions of the force may be required to defend the important types of terrain described below:

- Man-made features such as ports, key logistic installations, roads, railroads, water pumping stations, airfields, and wells.
- Natural features, such as mountain passes, or dominating ground, such as Mount Hermon on the border of Syria and Israel, or the Solium escarpment near the sea between Libya and Egypt.
- Key or decisive terrain that need not necessarily be a major feature, but one whose loss will inhibit the force in some manner. For example, the loss of terrain relatively close to a lodgement area may hinder the planned rate of buildup.

With the exception of the above cases, the retention of desert terrain normally makes little difference to the final outcome of battle. This does not mean that a commander has complete discretion to move his force wherever and whenever he wishes, as this movement will affect the dispositions of other US forces or allies. It means that possession of terrain is less important than the destruction of enemy forces. Although it will be necessary to dominate certain terrain or retain freedom to maneuver in large areas of the desert, there is no more sense in permanently occupying such areas than occupying a patch of sea. Assuming equal equipment capabilities for both opposing forces, the critical factor in defense will be the force ratios involved and the state of morale and training of the opposing forces.

A defense using aggressive maneuver at all levels is the best way to destroy large numbers of enemy without being destroyed in the process. If the defending force fails to remain mobile and active, the enemy will easily outflank it and strike directly at vital targets, such as the lodgement area. It is almost certain that one flank or the other will be open as were the south flanks of the British and German forces in Egypt and Libya in 1940-43. Since it will not be possible to maintain an unbroken line between strategic obstacles, air and ground security forces must be positioned in width and depth to guard against an enemy trying to outflank the defender.

Obstacles, both natural and artificial, are used to slow, contain, or isolate enemy units in order to defeat and destroy his units one at a time. Forward units block the enemy and canalize him into one or two avenues where he can be engaged from the flank. A reserve can then counterattack to destroy any remaining enemy.

Mutual support is normally a factor of time rather than weapon range due to the large areas to be covered. Gaps in initial positions may have to be accepted between and within task forces; although the ideal is to site units in such a manner that forces in at least two positions can engage an enemy maneuvering on any one of them. This greatly reduces any possibility of defeat in detail. When gaps exist they must be kept under surveillance. The defensive plan must include provisions for maneuvering to fire on any part of a gap before the enemy can move through it. A unit's area of responsibility must be defined by higher headquarters and should be clearly identifiable on the ground, which, due to the absence of significant terrain features, may require marking by artificial means.

FUNDAMENTALS OF THE DEFENSE

The fundamentals of defense are described in FM 100-5/FMFM 6-1 and in doctrinal manuals appropriate to each level of command. The following paragraphs discuss some points to remember in desert operations as they apply to the fundamentals of defense.

Reconnaissance and security units and force surveillance systems must focus on-

- What is the enemy's short-term objective?
- What are the enemy's avenues of approach, and what force is employed on each of them?
- Are the apparent movements real or feints?

As soon as these questions have been answered the commander will be able to maneuver to destroy the enemy. Until they are confirmed he can do nothing more than react to enemy initiatives, This is dangerous in any circumstance and doubly so in the desert as the side with the greatest potential for maneuver is more likely to win.

Direct-fire weapons must be used to their maximum effective range both by day and night. Limitations in night-vision equipment cannot be allowed to reduce depth or frontages; so plans for field artillery or mortar illumination are made for defense during limited visibility.

It is essential that all elements of a force retain their tactical mobility and efficient communications so that they can immediately react to changes in the commander's plans. Each individual weapon must be sited in a number of firing positions, even though vehicular movement may be exposed to air attack. Infantry fighting vehicles must remain in positions where they are concealed, capable of giving fire support to the dismounted squad, and available for immediate remounting.

Combined arms teams are essential to give the commander the capability he requires to fight the defensive battle. Defending forces orient on primary enemy approaches but units must also be prepared for attack from any other direction. Itis neither possible nor necessary to have maximum firepower in all directions, provided weapons can be moved to threatened areas before the enemy reaches them. Air cover or an air defense umbrella is necessary for a successful defense.

It is rare to find positions where any substantial part of the unit area of operations can be protected by natural obstacles. This require extensive use of artificial obstacles, depending on time, personnel, and combat service support available. Obstacles are used to divide the enemy force to improve local force ratios, and to slow the enemy's advance, thus permitting a flank attack. Conventional minefield must be clearly marked on the friendly side and recorded to avoid unnecessary losses if friendly forces later maneuver over the area.

STRONGPOINTS

Strongpoints are rare in desert warfare; however, they may be necessary to defend an oasis, mountain pass, or other key terrain essential to the defender's scheme of maneuver. When it is necessary to deny terrain to an enemy force, it is far better to initiate the defense well forward of the terrain feature, conduct the defense in depth, and destroy the enemy or force him to break off his attack before he reaches the critical feature.

In some cases the level of fortification and the deployment of the enemy may be a function of time, or

the enemy's intention and his understanding of what our forces are intending to do. The effectiveness of these strongpoints depends on the range of fires, the level of fortifications, and the decision of the opponent to attack them.

Deeply dug and well-prepared strongpoints surrounded by a minefield and having underground accommodations are usually used in the desert. Although these strongpoints may be neutralized by air or artillery fire and bypassed, eventually they will have to be assaulted. If they have been carefully sited and are well defended they can be quite effective. Variations of the strongpoint defense are used in rear operations. Combat service support units will use this method in perimeter defenses or base-cluster defenses. See Figure 3-8 for an example of a strongpoint and Figure 3-9 for an example of a strongpoint holding key terrain.

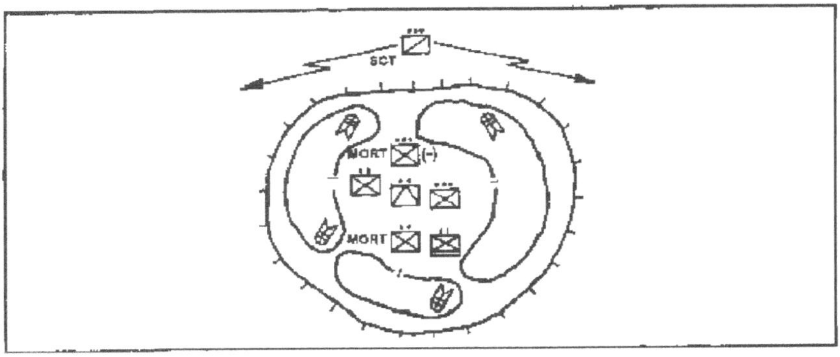

Figure 3-8, Strongpoint

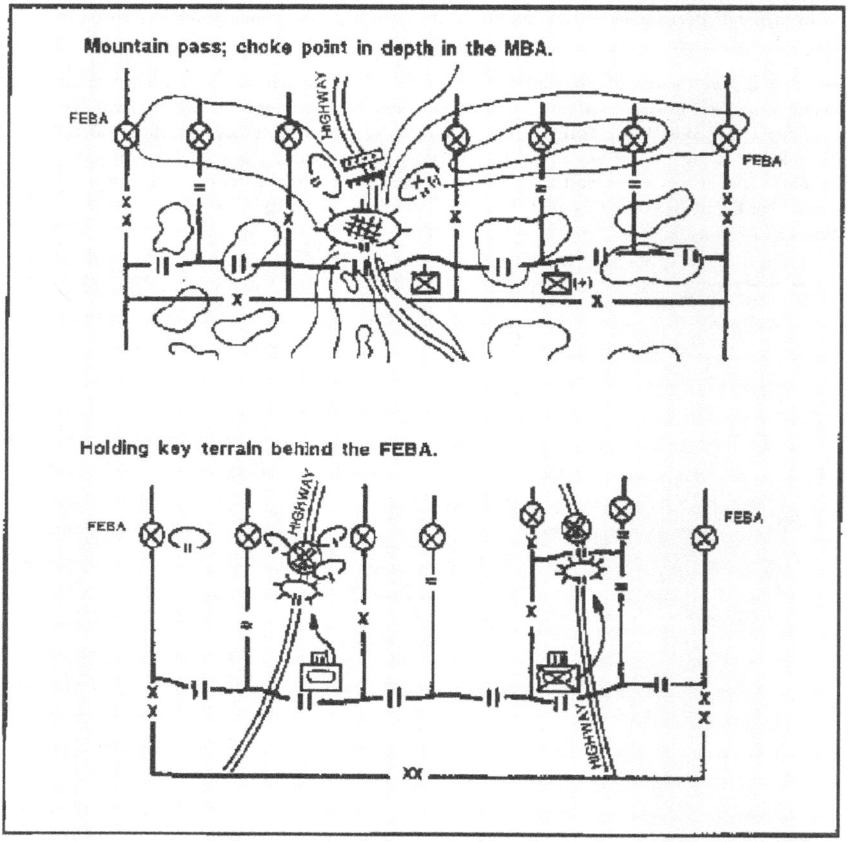

Figure 3-9, Strongpoints holding key terrain.

REVERSE SLOPE DEFENSE

The use of the reverse slope defense takes on added importance in the desert. Concealment is hard to achieve in the open desert. Detection of a unit's location invites both direct and indirect fires in abundance. The use of reverse slope positions will deny the enemy direct observation of positions until he is within the range of direct-fire weapons. Reverse slopes can even be found on seemingly flat desert floors; an intervisibility line will provide the reference for the establishment of engagement areas to support a reverse slope defense. A common misconception is that the desert is flat, when in fact, deserts are normally very uneven, with large breaks in the terrain.

Desert environments give special significance to the terrain aspect of METT-T. Commanders at all levels should place emphasis on the impact of desert terrain as it relates to the other factors of METT-T. The reverse slope defense in desert terrain warrants special considerations.

Direct-fire positions should be placed at the maximum effective ranges from the intervisibility line. This is where the enemy cannot see or engage a force with direct fire until he is within its engagement area. He can only deploy limited forces at a time. This allows the defender to mass fires on one portion of the enemy force at a time. The attacking force will have difficulty in observing and adjusting indirect fires. Obstacles may not be seen by the enemy until he is upon them and force him to breach under massed frees. Observation posts (OPs) positioned forward to see the advance of the enemy can influence the fight through indirect fires. The OPs can direct indirect fires on enemy forces that are slowed or stopped outside direct-fire ranges.

This defensive technique may be used in all defensive missions. Light infantry units use the reverse slope for protection against enemy long-range fires and to reduce the effects of massive indirect fires (artillery and close air support). The reverse slope defense brings the battle into the defender's weapons' ranges. Use of the reverse slope provides an opportunity to gain surprise.

The goal is to cause the enemy to commit his forces against the forward slope of the defense, resulting in his force attacking in an uncoordinated fashion across the crest. A reverse slope defense is organized on the portion of a terrain feature or slope that is masked from enemy direct fires and observation by the topographical crest, and extends rearward from the crest to maximize the range of the defender's weapon systems. See Figure 3-10 for an example of a reverse slope defense and Figure 3-11 for the organization of the reverse slope defense.

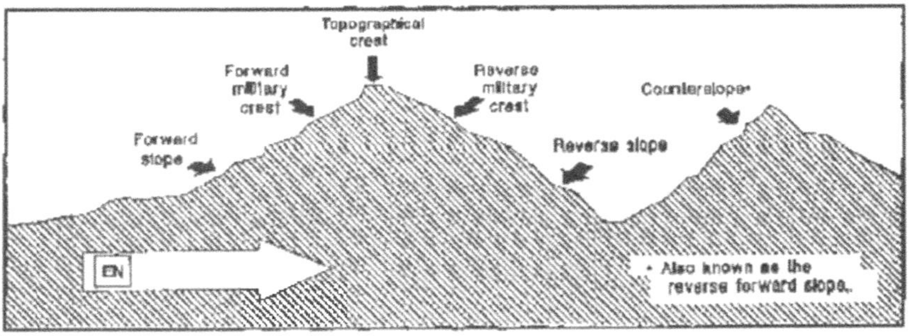

Figure 3-10. Reverse slope defense.

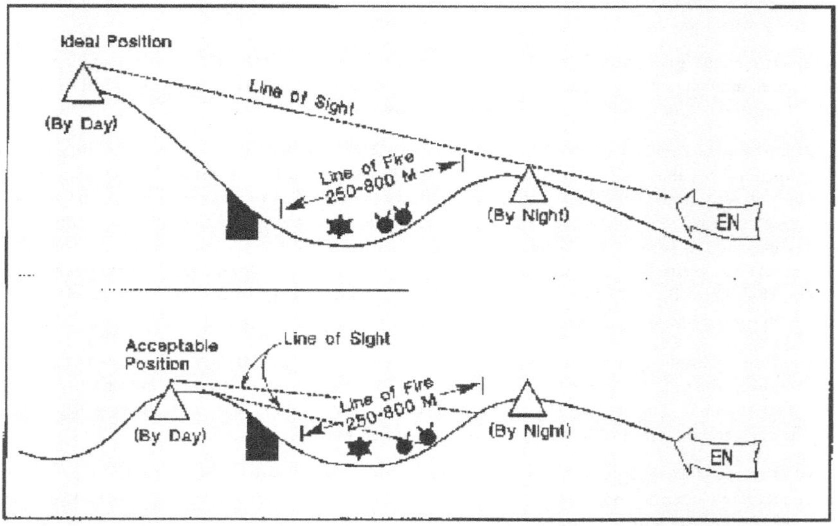

Figure 3-11. Organization of the reverse slope defence.

A disadvantage may be that the maximum ranges of weapon systems may not be employed due to the terrain available. The desert may be the best environment for the reverse slope defense. It may allow the use of weapons at maximum ranges as well as facilitating advantages. The following are advantages of a reverse slope defense:

- It hinders or prevents enemy observation of the defensive position.
- Attacking forces will not be able to receive direct-fire support from following forces.
- Enemy long-range antitank fires will be degraded.
- Attacking enemy forces will be silhouetted on the crest of the hill.
- Engineer work can be conducted away from direct-fire and observation from the enemy.

Reverse slope defense is not one concept, but a series of concepts that produce the potential for success. The concepts are-

- Pursue offensive opportunities through surprise and deceptive actions, with the intent of stealing the initiative, imposing the commander's will on the enemy, and breaking the enemy's morale.
- Afford the defender a variety of options in positioning his troops, with each option designed to draw the enemy into unfamiliar terrain.
- Enhance light infantry effectiveness and survivability.

A hasty or deliberate reverse slope defense may be considered when any of the following conditions exist:

- When the forward slope lacks cover and concealment, and effective enemy fire makes that position untenable.
- When the terrain on the reverse slope affords appreciably better fields of fire than those available on the forward slope.
- When it is desirable to avoid creating a distortion or dangerous salient in friendly lines by relying on forward slope positions.
- When it is essential to surprise and deceive the enemy as to the unit's true defensive positions or main effort.
- When seeking to gain protection from the enemy as he is massing fires.

DELAY OR WITHDRAWAL

When it is necessary to delay or withdraw, a desert offers many advantages to the defender. Long-range fields of fire allow engagements at maximum effective range of direct-fire weapon systems, and disengagement before the defender's position. However, dust clouds created by a moving force make it necessary to disengage under cover of smoke or darkness. Even a sandstorm can be used to the advantage. Field artillery, US Air Force fighter bombers, and attack helicopters can also be used to allow a ground maneuver unit to disengage and move rapidly to the next position.

When it is necessary to trade space for time, often a counterattack to destroy enemy advance units will do more good than trying to defend longer from an intermediate position.

Commanders at all levels should clearly understand the scheme of maneuver concept of the operation, and what it is they are expected to do, especially if communications should fail. Plans must include provisions for alternate means of communication. Routes should be clearly marked and reconnoitered to the maximum extent practical.

Due to the distances involved and constantly changing task organization, passage of lines is more difficult to coordinate and control. Pay extra attention to the identification of vehicles, routes of passage, signals, and coordination of movements.

Deception should be a part of all desert retrograde operations. The object of deception is to conceal the fact that a retrograde operation is taking place and that units are thinning out. Smoke and dummy positions can be used, false radio messages transmitted, and even dust clouds used to deceive the enemy.

ENVIRONMENTAL CONSIDERATIONS

In the desert it is necessary to modify the techniques of defense as described in doctrinal manuals applicable to each level of command and according to the mission, the fundamentals described in the preceding paragraph, and to the environmental considerations that are described in the following paragraphs.

Observation

The enemy will try to attack when the sun is low and behind him so as to dazzle the defender. The defender's observers must be as high as possible above the desert floor to see the advancing enemy as soon as possible.

Active light sources can be detected from great distances, especially during nights with low ambient light. Positive control of active light sources must be maintained until the battle is joined. Even then, the force equipped with passive devices will have the advantage over the force that is not equipped with these devices.

Heat from combat vehicles can give an enemy using thermal imagery devices a complete picture of the defensive scheme. So, combat vehicles should not prematurely occupy battle positions at night.

Sandstorms

Sandstorms may be used by the enemy to hide an offensive operation especially if the storm is blowing from the enemy's direction. When this is the case, units should immediately occupy their battle positions before the storm arrives. The unit should remain there until it ends, ready to fire and maneuver against the attacker after the storm abates. If vehicle patrolling is possible, a scout platoon or similar unit should cover all gaps, preferably moving in pairs, and on straight lines in view of navigational difficulties.

Terrain

From the point of view of a defending brigade or battalion task force commander, avenues of approach will often seem unlimited. Long-range observation must be maximized and scouts employed well forward to offset this problem. Radars should also be used extensively to provide early warning. It is necessary to identify the enemy's main effort early in order to move to concentrate.

Lack of concealment, especially from the aerial detection, prohibits units from occupying firing positions until just before engaging the enemy. Combat vehicles must displace immediately after engagement or risk destruction. Because of frequent displacement, mutes between battle positions should be reconnoitered and marked when possible, without revealing the scheme of defense. Smoke must be used frequently to conceal movement.

TACTICAL DECEPTION OPERATIONS

Analysis of desert operations from World War II to the present day indicates that tactical deception and surprise are clearly linked to the ability to move and mass forces during periods of limited visibility.

Operational planning should emphasize night movement of units. To minimize the problems of dust and to enhance deception, movement should be accomplished using multiple routes. Place priority on training to support this requirement. Associated with night movement is the requirement for night passage through lanes in minefield and forward passage through friendly forces.

In every modem desert war, deception has played a major role. The lack of concealment leads commanders to believe that with a reasonable reconnaissance effort they can gain an accurate picture of the enemy's dispositions. Reconnaissance by German, British, Israeli, Egyptian, and Syrian forces in modern desert warfare has been sufficient to detect the presence of combat forces in the desert. Deception has been successfully used in each of the modem desert conflicts to mislead commanders.

Since the desert environment makes it difficult to hide forces, the alternative is to make them look like something else-trucks and plywood made to look like tanks, and tanks made to look like trucks.

The movement of personnel and equipment and the placement of logistic support installations are normally indicators of a force's intent. The movement of empty boxes or pallets of ammunition and the establishment of fuel storage areas with real or dummy assets can deceive the enemy as to planned offensive actions. Use minimal actual transportation assets and make numerous, visible trips to simulate a large effort.

There are many examples of successful deception efforts by US forces from World War II In September 1944, the 43rd Cavalry Reconnaissance Squadron (Reinforced) occupied a 23-mile fronton the left flank of XX (US) Corps on the Metz Front. This squadron portrayed an armored division for several weeks and was so successful that the German Order of Battle Maps showed the 14th (US) Armored Division (AD) to be in the area. The 14th AD was not even in Europe at the time. Expertise in deception operations is critical to success.

Deception plays a key part in offensive operations and has two objectives: the first objective is to weaken the local defense by drawing reserves to another part of the battlefield. This may be done by making a small force seem larger than it is. The second objective is to conceal the avenue of approach and timing of the main attack. Some deception methods that can be used in offensive operations are-

- Using dummy units and installations.
- Using phony radio traffic.
- Using movement and suppressive fires in other areas timed to coincide with the real attack.
- Using small convoys to generate dust clouds.
- Filling ration boxes with sand and stacking them at landfills.
- Moving trucks into and out of the area giving it the appearance of being a storage facility or logistic base.
- Emulating damage to induce the enemy to leave important targets alone. For example, ragged patterns can be painted on the walls and roof of a building with tar and coal dust, and covers placed over them.
- Stacking debris nearby and wiring any unused portions for demolition. During an attack, covers are removed under cover of smoke generators, debris scattered, and demolitions blown. Subsequent enemy air photography will disclose a building that is too badly damaged to be used. Troops using the building after an attack must guard against heat emissions after dark and care must be taken to control electromagnetic emissions.
- Using phony minefield to simulate live minefields. For example, disturb the ground so that it appears that mines have been emplaced and mark boundaries with appropriate warnings.
- Making a real minefield to appear as phony or camouflaging it. For example, once a real minefield is settled, a wheel or a specially made circular wooden tank track marker can be run through the field, leaving track or tire marks to lure the enemy onto live mines. Antipersonnel mines should not be sown in such a field until the track marks have been laid. Another method is to leave gaps in the mechanically laid field, run vehicles through the gaps, and then close them with hand-laid mines without disturbing the track marks.
- Using decoys to confuse the enemy as to the strength of friendly forces and the unit's identity, or to conceal unit movement by being sited in a position after the real unit has moved. For more information on deception operations, see FM 90-2.

LONG-RANGE SURVEILLANCE OPERATIONS

Desert characteristics affecting LRS operations are: lack of water (a major problem), scarcity of vegetation, extensive sand areas, extreme temperature ranges, brilliant sunlight, and usually excellent observation. Movement using animals, vehicles, or by foot may be considered and is generally restricted to darkness. More training in land or air navigation and terrain orientation procedures may be necessary.

AIR ASSAULT OPERATIONS

An air assault task force provides commanders with truly unique capabilities. They can extend the battlefield, move, and rapidly concentrate combat power like no other forces.

An air assault task force uses the helicopter to move to and close with the enemy. Initial assault elements must be light and mobile. They are often separated from weapon systems, equipment, and materiel that provide protection and survivability on the battlefield. Thus, an air assault task force may be particularly vulnerable in a desert environment to enemy-

- Attack by aircraft and air defense weapon systems during the movement phase due to differences in desert effects on observation and fields of fire.
- Attacks (ground, air, artillery) during the loading and unloading phases and at other times when the infantry is not dug in.
- Small arms fire that presents a significant threat to helicopters.
- Artillery or other fires that may destroy helicopters and air assault forces during PZ (pickup zone) or LZ (landing zone) operations.

Air assault operations are discussed in greater detail in FM 90-4. Marine Corps assault support ensures the rapid buildup of combat power and facilitates the quick maneuver of ground forces. See FMFM 5-35 for more information.

AIRBORNE OPERATIONS

The airborne division is organized to rapidly deploy anywhere in the world. It is the only US Army division with a rapid, strategic, combined arms, forced-entry capability. It will most likely be the initial force deployed for contingency operations. It is ideally suited and primarily designed to seize, secure, and repair airfields in order to provide an airhead for follow-on forces and to delay, disrupt, and reduce enemy forces.

Airborne operations can be adversely affected by various environmental considerations of the desert. High winds, sandstorms, and heavy rainfall or thunderstorms may impact on mission accomplishment. Planning considerations for combat service support are complicated by the fact that airborne forces will be the first Army forces in an immature and austere desert theater. For more information on airborne operations, see FM 90-26.

MARINE AIR GROUND TASK FORCE OPERATIONS

Marine operating forces are organized for combat as Marine air ground task forces (MAGTFs) composed of command, ground combat, aviation combat, and combat service support elements. The MAGTFs are closely integrated combat forces capable of rapid response to any crisis or contingency. Their Naval/Marine Corps expeditionary nature makes them ideal for immature and austere environments as was seen in Operations Desert Shield/Storm (Southwest Asia) and Restore Hope (Somalia).

A variety of types of MAGTFs may be formed in support of national strategy and rapid crisis response. The Marine expeditionary force (MEF) is the Corps' principal organization for combat and peacetime readiness, and is formed from the legislated division and aircraft wing teams. These MEFs provide a reservoir of integrated combined arms combat power that can be task organized to simultaneously execute a wide range of global missions. The MAGTFs are mission tailored and range in size from very powerful MEFs, capable of prosecuting operational campaigns against the most capable potential threat through rapidly deployable and employable Marine expeditionary units (MEUs), to small special purpose forces (SPMAGTFs) formed for specific missions or crises. In the early moments of Operation Desert Shield, a MEF provided the nation a powerful combined arms combat force to stand against aggression, while US forces and equipment and supplies were being assembled in Southwest Asia. For more information on MAGTF operations, see FMFRP 2-12.

MARITIME PRE-POSITIONING FORCE OPERATIONS

The maritime pre-positioning force (MPF) gives the nation an added dimension in mobility, readiness, and global responsiveness. The MPF program involves 13 ships, organized in three squadrons. Maritime pre-positioning squadron one (MPSRON-1) operates in the eastern Atlantic Ocean, MPSRON-2 in the Indian Ocean, and MPSRON-3 in the western Pacific. The MPF, when called upon, provides equipment and 30 (days of supplies for a 16,000-man Marine expeditionary brigade (MEB). The MEB's personnel and selected equipment can be airlifted quickly using roughly 250 airlift sorties to an objective area to join with its equipment at a secure site. Equipment and supplies can also be selectively off-loaded to support smaller MAGTFs. During Operation Desert Shield/Storm, all three MPFs were off-loaded in Southwest Asia providing immediate support to deploying forces. During Restore Hope, one MPF supported operations. For more information on MPF operations, see FMFM 1-5.

ARMY SPECIAL OPERATIONS FORCES

To meet our nation's global commitments, the Army maintains a balanced force of armored, light, special operations, and support forces for use across the operational continuum. Army special operation forces (ARSOF) are an integral part of the total Army force. ARSOF have five elements: special forces, rangers, Army special operations aviation (ARSOA), PSYOP, and civil affairs. These forces offer significant capabilities to the desert theater of operations. Details of ARSOF capabilities are discussed in FM 100-25.

MILITARY OPERATIONS ON URBANIZED TERRAIN (MOUT)

Because there are few man-made features throughout the expanse of the desert, those that do exist can become important, perhaps even key. Key terrain in the desert can be any man-made feature. Settlements (where a logistics base may be established), road junctions, shelters, and airfields, all become important, simply because they are so few in number. Growing villages and settlements straddle these lines of communication, and small villages may exist near water sources and other key terrain. It may be necessary to conduct MOUT operations to control these areas. In areas involved with Desert Storm, paved roads and even dirt roads were considered key terrain for both high-speed movement and for providing clearly defined directions and locations. Commanders must be prepared to fight on terrain that is constantly being modified by man. More information on conducting MOUT can be found in FM 90-10.

CHAPTER 4

COMBAT SERVICE SUPPORT

There is a lack of every kind of resource in the desert, especially of the sophisticated infrastructure of ports and railways for their high capacity for moving combat supplies. Logistical support is always a challenge, and an arid environment burdens all types-supply, aviation, communications, and maintenance. Commanders must be sensitive to the constraints, and those providing support must work to overcome them. A unit's tactical effectiveness in the desert depends to a large degree on the combat service support available. Equally, its vulnerability lies in its exposed lines of communications and the immobility of its bases of supply and support.

Inherent to the success of any tactical operation is continuous, sound, logistical planning for adequate supply, medical, and maintenance support. This is especially important in the desert because the greater distances used in maneuver and deployment complicate supply procedures. Other reasons are the shortage of locally available water and the increased maintenance requirements due to sand and dust damage. The effects of the environment on equipment are severe, requiring increased levels of support to maintain a standard level of efficiency. The extended supply lines required for expanded frontages call for special considerations and procedures to ensure adequate and timely supplies arrive to sustain combat in the desert.

US forces in the desert operate at the end of a long, perhaps tenuous line of communication. Cargo space must not be wasted to provide all the comforts of home. A significant difference in living standards between rear area support personnel and those in forward combat areas must be avoided since this can affect morale and weaken the ability of combat units to resist psychological warfare. Transportation priority must be given to minimum essential materials and the support base should be austere.

Section I. Base Development Plan

US forces deploying for operations in a desert environment should expect to begin operations from a lodgement area. When this is the case, it is necessary for the headquarters deploying the force to prepare a detailed base development plan. How the plan is developed will depend on a number of factors that are described below:

- The mission and size of the force. The size of the force depends on its mission and the operations it is expected to conduct. The size of US forces deployed for desert operations could vary from a small force conducting a show of force, to a joint task force capable of full-scale operations.
- Security of the lodgement area. A lodgement area will probably be secured by allied forces or US Marines before deployment of US Army forces into the operational area. However, it may be necessary to use US Army forces, either air-dropped or air-landed, to secure a lodgement area.

- Transportation of US forces into the lodgement area. It is probable that initial forces will be transported by air and follow-on forces by sea. Another possibility is that initial forces will be transported by sea with follow-on personnel being transported by air.
- Strategic lines of communication (LOC). The initial strategic LOC will probably be an air LOC. However, at some point in the operation a sea LOC will be established to convey the bulk of the supplies, supplemented by an air LOC to haul time-sensitive items.

- Theater lines of communication. Lines of communication within the operations area should be analyzed before selecting the lodgement area. The analysis should include ports available, airfields throughout the operational area, road nets, and railroads. It may be necessary to stage engineer construction units into the operational area early to improve existing facilities and LOCs or to construct new ones. In a single or multi-corps theater, a theater army headquarters provides overall management of CSS operations. It establishes priorities, assigns missions, and allocates resources in accordance with the theater army commander's concept of operations.
- Local resources. These are extremely important as they will affect logistics planning. Typical information about resources in the operational area that should be obtained before base development planning includes-

  -Airfields.

  -Water sources.

  -Fresh rations.

  -Labor supply.

  -Construction materials and available equipment.

  -Material-handling equipment at ports and airfields.

  -Local hospitals, maintenance capability, and storage sites.

  -Local power supply to include types and equipment.

  -Railroad rolling stock and gauge of tracks in local areas.

After consideration of the factors listed above, the lodgement area is selected Ideally, a lodgement area should have a deep water port and airfield suitable for heavy strategic airlift, located at the end of an adequate road or rail system suitable for an intratheater LOC. Once the lodgement area has been selected, then LOC-port units can be specially tailored for early deployment to the operational area.

Section II. Theater Support

THEATER HEADQUARTERS COMMAND

Should a friendly nation ask for military assistance either following an invasion or in anticipation of one, the first combative units to arrive in the theater will probably be a force designed to secure entry points. The assembly and movement of armor and mechanized forces will take time, time that the logistics staffs can put to good use preparing for the reception of these forces. Elements of the theater staff can assess the assets and facilities available in the following areas and make an estimate of the work, labor, equipment, and other resources required to support the buildup of the force.

Logistics Reconnaissance

Port capacity, cranes and off-loading equipment, storage facilities, the availability of stevedores, and local shipping assets at the main port facilities should be determined. Contract and requisition of host nation support assets can be arranged by initial theater staff personnel. The theater staff should also determine airfield facilities and capacities that are available. Railway assets that can be made available for logistics support and the movement of troops should also be determined.

Local Resources

The host nation in a Middle Eastern theater will probably be an oil producer and may be able to provide the bulk of our fuel requirements. However, if the host nation exports its oil in bulk crude, it may have only a limited refinery capacity to meet a local domestic demand for diesel, kerosene, high-octane fuel, and lubricants.

Fresh Rations

The host nation may be able to produce a limited supply of fresh rations but the bulk may have to be imported from neighboring countries. If there is arable land within the host nation, it may be possible to start farms to relieve the burden on the local economy and on that of neighboring friendly states.

Hospitals

The extent to which hospitals in the host nation can accept long-term patients until our own base hospitals can be established is also an issue of concern.

Power Supply

The share of the host nation's electric power supplies which can be offered to our forces must be determined. Initially, there is unlikely to be any shortage in power supply, but as a large force (including other allies) builds up there maybe a generating capacity problem. At the beginning of the campaign the difficulty may be in distributing electricity where it is required throughout the theater. It may be necessary to construct overhead or underground cables together with transformers.

Transient Camps

Sites should be located for transient camps for troops arriving in the theater and convalescent camps for the recovery of the sick and wounded.

Local Currency

Local currency may also be required for extended operations, both to pay troops and to locally purchase supplies.

Maps

Maps are a critical item that maybe more readily available through local survey teams or oil companies.

Interpreters

Interpreters will be required to communicate with the host-nation troops, contractors, and labor forces.

Stockpiles

Higher stock levels are required in the desert due to the following factors: use of the limited life of many perishable items in a harsh environment; the enormous distances stockpiles must be lifted and over inadequate transport systems; the loss of supplies due to sudden changes in the fortunes of war, and the time it takes to replace items from the US. The levels of each commodity to be held in theater and the proportion of the totals will be decided during the staff planning process. The distribution will depend on the tactical situation and the vulnerability of the lines of communication to enemy action.

Expenditure Rates

While the requirement for rations and water remain relatively constant, the expenditure of fuel and ammunition will vary far more, not just because of the fluctuation between quiet periods and intense operations, but because of the environment. The amount of driving in soft sand and the longer distances to be traversed combine to increase consumption beyond central European rates. Similarly, the expenditure of tank and artillery ammunition may be increased because of the open terrain.

Host-Nation Support

In a desert environment where resources, CSS personnel, and equipment are limited, the use of host-nation support assets can be vital to the success of an operation. Host-nation support assists in the accomplishment of missions and functions in support of US forces and enhances their capability to perform their wartime role.

All forms of peacetime transition to wartime, and wartime host-nation support should be included in the planning process.

Host-nation support includes-

- Government agency support such as police, fire companies, and border patrols, may be available to support US forces.
- Contractor support such as supplies and services, including laundry, bath, bakery, transportation, labor, and construction.
- Host-nation civilians may be able to provide needed skills for laborers, stevedores, truck drivers, managers, and technicians.
- Host-nation military units may provide traffic control, convoy escort, installation security, cargo and troop transport, POL storage and distribution, and rear operations.
- Host-nation facilities may be contracted and used for hospitals, headquarters, billets, maintenance shops, or other activities.

- Functional or area support maybe provided in the form of rail operations, convoy scheduling, air traffic control, and harbor pilot services.
- Services may be provided by the host nation for gymnasiums, recreation facilities, and other morale and welfare demands.
- Supplies and equipment needed for missions may be acquired locally, precluding or reducing materiel shipments from CONUS.

Section III. Corps Support Command

An Army corps support command (COSCOM) deploying to support desert operations must be carefully tailored to meet the needs of combat forces operating in a harsh environment. Requirements for long-haul truck companies, engineer construction battalions, water production units, and LOC-port units previously described, must be carefully weighed. A shortfall of these units could significantly impair combat operations. Organization of the COSCOM should be planned based on the factors described in the previous paragraphs, with particular attention given to-

- Number of troops to be supported.
- Quantity and types of equipment to be maintained.
- Tonnage to be handled.
- Available local resources and labor force.
- **Types** of units to be deployed to the theater of operations.

The organization of the COSCOM and a description of its tasks are provided in FM 63-3J. Initial corps forces entering the theater can be supported by a forward support battalion (FSB) of a division support command and a corps support battalion (CSB) of a corps. Once initial forces have arrived in the theater, additional tailored elements from the COSCOM must immediately follow, or even arrive first to minimize the requirement for the DISCOM cooperate such activities as ports or airheads.

Section IV. Division Support Command

As previously mentioned, combat service support units are high-priority targets for any desert enemy. In most cases, Army division support command (DISCOM) units will not be able to provide for their own security, considering the many ways in which they could be attacked. Air defense protection must be provided. It may even be necessary to provide a maneuver unit or additional MP units to secure DISCOM elements. Nearby maneuver units can also be designated to move to their defense-attack helicopters are especially suited for this purpose-and on-call fires should be planned by nearby field artillery units. Any pipelines in the division area must also be secured by any means at hand. Observation helicopters can be used to patrol pipelines.

Stocks should be kept as mobile as possible in the event rapid displacement is necessary. Stockpiling off vehicles must be held to a minimum, as should stockage levels. To the extent practical, supplies located forward of the division support area should be stored aboard vehicles to minimize the Possibility of having to leave them behind. For this purpose, a force operating in the desert should be augmented with additional transportation assets.

DISCOM organizations of the lead divisions in austere and immature theaters may be called upon to establish forward logistic bases. In these situations division assets may have to assume other support or transportation responsibilities temporarily until area support groups can establish support operations and transportation.

Section V. Combat Service Support Element

The Marine Corps' combat service support element (CSSE) is a task organized service support element of the MAGTF. Its composition is based on many factors, to include-

- MAGTF size (MEF, MEB, MEU, or SPMAGTF).
- MAGTF mission.
- Type of operation.
- Area of operation.

The considerations listed under Sections III and IV are also true for a CSSE. For more information on CSSE operations, see FMFM 4-1.

Section VI. Support Operations

Listed below are some of the factors that make support operations complicated.

SUPPLY

Consumption rates must often be developed after the force has operated for some time in the area. Water has to be found, purified, stored, and transported.

MAINTENANCE

There will be a greater demand for such items as filters, oils, and lubricants. More Class IX stores are required than normal, and the work load on maintenance units is much greater. Supply items and spare parts should be packed or wrapped as if to be air and water tight to prevent blowing sand from contaminating or damaging them. All echelons that request supplies and repair parts should be using the same or compatible equipment for requisitioning, with alternative means in place as a backup so there is little or no slow down in the reorder process.

CONCEALMENT

It is difficult to conceal trains areas. However, trains areas must be concealed to the best extent possible. These are soft targets in any environment and are high-priority enemy targets as their destruction (especially water, HETTS, and fuel supplies) effectively cripples the force.

DISTRIBUTION

Maneuver units may be farther apart, both in width and depth, than in temperate environments. They move more frequently and faster. Lines of communication are longer. Terrain away from the main supply mute (MSR) maybe such that it is only trafficked by cross-country vehicles, and then only with reduced payloads. Lack of significant terrain features may increase navigational problems, requiring local guides.

COMBAT SERVICE SUPPORT PLANNING

The commander's intent and METT-T analysis must dictate the CSS plan to support the tactical mission. However, CSS planners must not become locked into rigid CSS plans. The situation will dictate how trains are configured, echeloned, and controlled. Commanders and their staffs must use a logical and fast means to evaluate the battlefield and reach decisions. The military decision-making process described in FM 101-5 provides the framework within which the commander and staff interact to arrive at and execute a decision. Battlefield support must be planned to satisfy requirements during the following operational phases:

- Prior to D-day (before).
- Commitment to battle (during).
- Future mission (after).

All areas of CSS (man, arm, fuel, fix, move, and protect) must be considered during each operational phase to ensure an integrated, responsive plan of support. Support requirements must be projected and plans developed to satisfy these projected requirements. Supporting CSS plans should be as detailed as planning time permits.

CSS commanders and planners must thoroughly understand the tactical mission and plans and the commander's intent. They must know-

- What each of the supported elements will be doing.
- When they will do it.
- How they will do it.
- Where they will do it.
- What the priority of support is.
- Density of personnel and equipment being supported.

After analyzing the concept of the operation, CSS commanders and planners must be able to accurately predict support requirements. They must determine-

- What type of support is required.
- What quantities of support are required.
- The operational commander's priorities, by type and unit.

Using the support requirement of the tactical plan as a base, the support capabilities of the CSS structure are assessed. The staff must determine-

- What CSS resources are available (organic, lateral, and higher headquarters).
- Where the CSS resources are.
- When CSS resources will be available to maneuver units.
- How the FSB will make these resources available.

Based on this information, the staff must then develop support plans that apply resources against requirements in a manner that results in the most responsive support possible. Communications links must be established and maintained.

Orders that clearly describe tasks to be accomplished must be issued. Continuous follow-up must ensure tasks are being accomplished as planned.

CSS functions should be performed as far forward as the tactical situation and available resources will permit. They should be performed at or close to the site where the weapon system is located to lessen evacuation requirements. Support must be continuous, using immediately available assets. This will involve bringing ammunition, fuels, parts, end items, maintenance personnel, and occasionally replacement crews or individuals, to the forward elements such as battalion field trains, combat trains, and equipment downsites. Planning and execution emphasize the concept of providing support to forces in the forward areas.

CSS planners must know priorities for support. This is necessary to ensure that units with the highest tactical priority receive required support first. The commander and his staff provide mission directives, determine CSS requirements, and establish priorities within the unit.

Section VII. Security of Supply Routes

Long lines of communications require convoys from the support base to the combat forces, and convoys are subject to air attacks (as learned during World War II when convoys from Casablanca to Al Guettar Tunisia, were frequently targeted by Luftwaffe raids), Enemy ambushes on main supply routes (MSR) are always a threat in desert operations. Enemy patrols may also place nuisance mines on routes, especially at critical points such as defiles. Actions must be taken to minimize the threat to supply routes.

The MSR will be considerably longer in the desert. The logistical assets utilizing the MSR are extremely vulnerable and must be protected. It maybe necessary to allocate maneuver forces to maintain an open and relatively safe MSR. This will allow supplies to be pushed forward and casualties to be moved to the rear. Coordination must be accomplished between maneuver elements as to where responsibility will end and begin on the MSR during each phase of an operation. The MSR requires constant patrolling to ensure safest operations and continuance of supplies to the maneuver forces. Marking of the MSR facilitates security. Different techniques for marking MSRs range from chemical lights, to spray paint, to signs. MPs can also position themselves along the MSR to help guide units.

Section VIII. Supply

Time and distance factors developed on different terrain by experience are of little value in the desert. The absence of roads in forward areas, navigation problems, vulnerability of trains and supply installations to attack by ground forces or aircraft, sandstorms, and wide dispersion, all require a different appreciation of time for resupply operations.

CLASSES OF SUPPLY

Requirements for supplies vary from that of temperate climates according to the classification of supply. Differences that may be expected in any desert are described in the following paragraphs.

Class I

Until the theater is fully developed and ration requests can be implemented, ensure enough MRE rations for three to five days are stored on combat vehicles. Meals from this combat load are eaten only when daily Class I resupply cannot be accomplished. Frequency of unit feeding and use of A or B rations depends on the tactical situations, If possible, troops should receive at least one hot meal per day. Hot rations should be packed in platoon-size portions rather than consolidating company-size packages.

It is critical to plan for the cooling of water supplies. Troops will drink any potable water available to them; however, they would prefer chilled water. Commanders and staffs must plan for water coding systems, ice, and individual soldier/marine field-expedient devices. Troops fighting in the desert will likely be wearing the battle dress overgarment and body armor. This fact will impact on the planning for water consumption.

The key to having enough water in a task force conducting desert operations is the capacity of that force to store and transport it. Current water trailers are inadequate. The potential for water consumption is high when you consider personal use and consumption, decontamination, medical needs, messing operations, and maintenance uses. Possible solutions include converting fuel tankers to water tankers, the use of blivets, and local purchase of civilian water-holding tanks through host-nation support. Water is vital, yet local supplies may be scarce or nonexistent in a desert combat zone. If water is plentiful, as it is in mess around Tripoli and Benghasi in Libya, water supply should not be a problem, provided that normal water supply procedures are followed.

This paragraph deals with situations where local water is difficult to obtain. All units must maintain a continuous watch for possible sites such as oases, dry wells, dry water courses, open water (even marshes), or captured enemy dumps. These should be reported to the next higher headquarters, giving the location and quantity and flow, if possible. It is not the responsibility of these units to test the water for potability, which could be dangerous for untrained personnel. This task should be left to specialists.

Since distances between water points may be long, it may be desirable to augment the division with additional 5,000- and 2,500-gallon bulk water tankers, processed to haul water.

Priorities for water use should be established. See Appendix G for a suggested list of water priorities and additional information concerning drinking water needs, water requirements, and water heating rates. If vehicle decontamination is necessary, it will take a high priority. NonPotable water should be used for this task.

Class II

There is little change in Class II consumption. However, clothing variations, from tropical clothing to sweaters and sleeping bags, must be anticipated Requirements for items such as neck scarves and canteens will be increased as well as those for hand tools, since tools tend to get lost in the sand.

Class ill

There is a marked increase in oils and lubricants used in preventive maintenance however, the actual quantities depend on operating conditions. Some types of desert terrain can lead to greatly increased fuel consumption per mile moved or hours that equipment is used. Use of cans or fuel bladders in certain circumstances should also be considered as they allow fuel to be spread more evenly among cargo vehicles since a loaded fuel tanker's cross-country capability may be degraded in desert sand. HEMTTS may be a suitable replacement vehicle to solve the cross-country mobility problem.

Antifreeze requirements remain roughly the same as in temperate climates as antifreeze increases the boiling point of coolant and decreases wear on liquid-cooled engines. Various oils and lubricants are required in smaller user containers. This assists in preventing sand from contaminating larger containers since they would have to be moved from site to site, and opened and closed numerous times.

Class IV

The requirement for Class IV stores can be significantly more than in other theaters, and consumption of some items such as sandbags is greatly increased. Maximum use must be made of local materials. An engineer reconnaissance unit should be present in the theater from the initial buildup to establish what resources are available. All possible Class IV should be carried and incorporated into vehicle load plans when deploying. The construction of airstrips, minor port facilities, and rehabilitation of major port facilities and railways, are all engineer missions of particular importance in desert warfare.

Class V

Due to excellent firing conditions, and the need for extensive suppressive fires, ammunition consumption may be high. It may be necessary to restrict firing of certain types of ammunition once they have reached predesignated levels unless command approval is obtained. Battalion task force trains should contain a one-day supply of ammunition and missiles for all vehicles in the task force. Ammunition should be divided between combat and field trains when trains are echeloned. Units should keep ammunition as packaged until it can be uploaded on combat vehicles. This will protect the ammunition from sand that could cause weapon malfunction.

If artificial obstacles are to be employed, considerable quantities of mines will be required as minefield must be long and deep to be effective. Since extensive minefield will be preplanned, relatively few antitank mines need to be held in ammunition supply points forward of the division support area. When required, the quantities needed should be moved as close to minefield locations as possible. Only mines necessary to replenish unit basic loads used for local defense need be stocked forward of the division.

Class VI

The demand for Class VI supplies, especially beverages, is high. They are not, however, essential and if transportation is limited they are given a low priority, especially if refrigeration space is certain to be in short supply. Sundries packs can also be used.

Class VII

The demand for Class VII supplies depends greatly on maneuver and the intensity of the battle. The only variation that can be forecast is for refrigeration equipment, especially if it is necessary to move deceased personnel to the United States for burial.

many of the deserts being in predominantly Muslim cultures, religious support may be affected and should be a consideration prior to deployment.

Section XI. Health Services

Medical unit requirements for desert operations are essentially the same as for temperate climates. It is essential that each brigade has an environmental sanitation team attached. When planning for medical support the following factors should be considered:

Increased dispersion and large areas over which battles are fought increases vehicle evacuation time. This problem can be further complicated if the enemy does not recognize the protection of the Red Cross, thereby inhibiting air evacuation within the range of enemy air defense weapons. The importance of units having trained combat lifesavers is critical to overcoming this. The reduction of the number of deaths due to slow evacuation time can be directly affected by the combat lifesavers available. One combat lifesaver per combat vehicle is an adequate number.

The comparatively long distances between units may limit the availability of medical aidmen to adequately support combat troops. Reinforcements may be required from the division medical battalion or from supporting corps level medical units. Augmentation should include vehicles as well as personnel.

The incidence of illness from heat injuries and diseases are higher than in temperate climates. Fevers, diarrhea, and vomiting, for example, cause loss of water and salt, which can culminate in heat illnesses. Cold weather injuries can also occur during a desert winter.

The mobility required of maneuver units will be inhibited if movement of any part of these units, including trains, is restricted by having to hold a number of casualties; therefore, the wounded and sick must be evacuated immediately.

In order to properly treat patients, all medical treatment facilities should be provided additional supplies of water. Medical personnel at all levels must assist tactical commanders in preventing or reducing heat casualties within their units.

Divisional medical units should be augmented with extra field ambulances from corps units. In an emergency, empty cargo trucks moving to the rear can be used for medical evacuation.

Evacuation from the battalion combat trains back to the brigade ambulance exchange point (AXP) or clearing station, will be performed by ground or air transportation. METT-T, availability of equipment, and the patient's condition, will be the determining factors on what method of transportation will be utilized.

The effects of nuclear weapons can be expected to be greater in desert terrain. Introduction of nuclear weapons by the enemy will greatly increase casualties and severely strain available medical resources. The same effects can be expected if the enemy introduces chemical weapons against unprepared troops.

Section XII. Naval and Air Force Assistance

NAVAL

During the initial stages of an operation it may be necessary to request logistic support from the US Navy. Ships, with the exception of a few special types, are neither designed nor equipped to give logistic support to ground forces. Limited support may be available if it is adequately coordinated in advance. A cruiser, for example, may have more than 20,000 gallons of water per day available beyond the requirements of the crew. Limited supplies of items such as bread may be available. Limited surgical and medical assistance may also be available.

AIR FORCE

The military air command (MAC) provides tactical airlift in support of the force. Air Force assistance must be coordinated with MAC to deliver personnel, supplies, and equipment forward to brigades and farther forward when necessary. Delivery is made by the most suitable means available, air landing, extraction, or airdrop. MAC also makes aircraft available for rearward movement of wounded persons or prisoners of war.

Section XIII. Other Combat Service Support Issues

MORTUARY AFFAIRS

The mortuary affairs program provides peacetime and wartime support to search, recover, identify, evacuate, and, when required, temporarily inter, disinter and re-inter deceased US military or civilian personnel, and allied and enemy dead. In addition, the program provides support to collect, inventory, store, and process personal effects (PE) of deceased, missing, captured, and medically evacuated US personnel and deceased allied and enemy personnel.

The goal of the mortuary affairs program is to search, recover, identify, and evacuate the remains of US military and certain civilian personnel from the theater of operations as long as feasible using available US Air Force aircraft.

The longer lines of communication required in desert environments may affect the evacuation of the deceased and therefore must be a planning consideration. Transport remains in palletized transfer cases when the tactical and logistical situation permits.

When the situation prohibits immediate evacuation, remains may have to be temporarily interred within the theater. When possible, the use of temporary cemeteries will be confined to echelons above corps. However, emergency war burial (mass burial) sites, as authorized by the theater commander, may be required as far forward as the brigade area. Desert environmental factors should be considered when establishing temporary interment sites, as the desert environment can significantly affect burial sites (strong winds, flash flooding and so forth. Remains will be buried with their personal effects in these burial sites to assist in the identification of remains when they are disinterred.

This concept will permit the use of host nation support (HNS) to dig and fill mass burial sites. Host nation support laborers will not actually handle or process remains or personal effects of US personnel, but will generally provide labor for interment operations. The mortuary affairs company commander has overall technical responsibility for the layout and survey accuracy of the cemetery. During World War II HNS/EPW labor was used successfully by mortuary affairs units.

CAPTURED MATERIEL AND PERSONNEL

In a desert theater of operations, where resources are scarce to begin with, the innovative use of captured materiel can be critical. This materiel can contribute to the retention of momentum by maneuver forces and decreases the need to consume our own supply stocks and to transport them to using units. Obvious sources are captured or overrun enemy fuel supply points, and materiel which may be used for barrier and fortification construction. Food and medical supplies may be used to feed and treat EPWs and civilians. Commanders and staffs must have a workable plan for handling EPWs. These potentially overwhelming requirements include health services, transportation, security forces, and so forth.

CLOTHING EXCHANGE AND BATH SERVICES

Clothing exchange and bath (CEB) services are provided by the supply and service company, when augmented. CEB services are requested through the brigade S4. The request must specify the location of the unit making the request, desired time for service, and range of clothing sizes for unit members. The requesting unit must be prepared to assist troops in setting up the CEB point.

NOTE: These activities may not be possible in the early stages of an operation, or not at all in forward areas. Additional effort will be required to provide these services in the desert.

APPENDIX A

DESERT COUNTRIES OF THE WORLD

This appendix provides brief descriptions of notable deserts of the world. Detailed information can be obtained from a terrain analysis from the Defense Intelligence Agency and the relevant volume of the DA Pamphlet 550-series. See Figure A-1 for desert locations of the world.

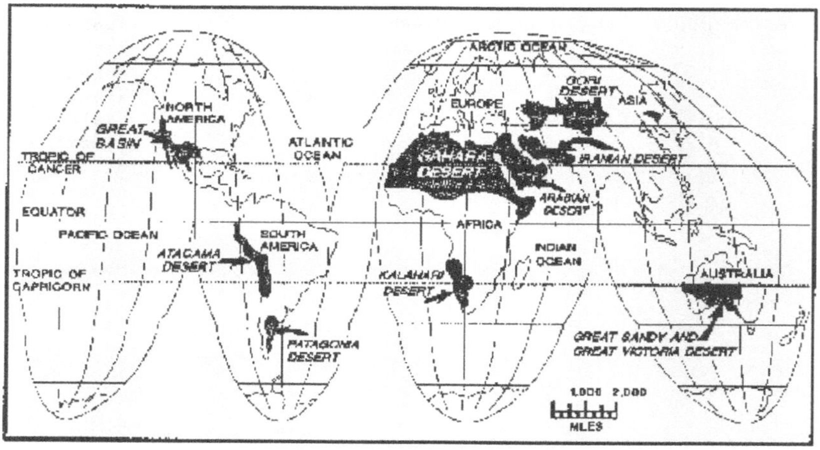

Figure A-1. Desens of the world

NOTABLE DESERTS OF THE WORLD

ARABIAN (EASTERN)	70,000 square miles	In Egypt between the Nile River and Red Sea, extending southward into the Sudan
ATACAMA	600 miles long	In northern Chile
CHIHUAHUA	140,000 square miles	In New Mexico, Texas, Arizona and Mexico

COLORADO	7,500 square miles	South of the Mojave desert, along the Colorado River in southern California
DEATH VALLEY	3,300 square miles	In eastern California and southwest Nevada
GIBSON	120,000 square miles	In the interior of western Australia
GOBI	500,000 square miles	In Mongolia and China
GREAT SANDY	150,000 square miles	In western Australia
GREAT VICTORIA	150,000 square miles	In western and southern Australia
KALAHARI	225,000 square miles	In southern Africa
KARA-KUM	120,000 square miles	In Turkmenia (former Soviet Union)
KAVIR	400 miles long	In central Iran
KYZLKUM	100,000 square miles	Kazakhstan and Uzbekistan (former Soviet Union)
LUT	20,000 square miles	In eastern Iran
MOJAVE	15,000 square miles	In southern California
NAFUD	40,000 square miles	Near Jawf in Saudi Arabia
NAMIB	800 miles long	Along the southwest coast of Africa
NUBIAN	100,000 square miles	In the Sahara in northeast Sudan
PAINTED DESERT	150 miles long In northern Arizona	In northern Arizona
SAHARA	3,500,000 square miles	In Northern Africa extending westward to the Atlantic (largest desert in the world)
SIMPSON	40,000 square miles	In central Australia
SONORAN	70,000 square miles	In southwest Arizona and southeast California
SYRIAN	100,000 square miles	In northern Saudi Arabia eastern Jordan, southern Syria,and western Iraq

Table 1-**1**. Desert counties of the world.

COUNTRY	AREA SQ MILES	POPULATION (MILLIONS)	LANGUAGE(S)	ADJACENT COUNTRIES
Algeria	919,487	24.2	Arabic French Berber	Morocco Western Sahara Mauritania Mali Niger Libya Tunisia
Bahrain	255	.480	Arabic English Farsi Urdu	Saudi Arabia Qatar
Egypt	386,660	53.3	Arabic English French Arabic	Libya Sudan Israel
Iran	636,293	51.9	Farsi Turkish Kurdish Arabic English	Iraq Turkey Former Soviet Union Afghanistan Pakistan
Iraq	167,924	17.6	Arabic Kurdish Assyrian Armenian	Syria Jordan Saudi Arabia Kuwait Iran Turkey
Israel	8,019	4.3	Hebrew Arabic English French	Egypt Jordan Syria Lebanon
Jordan	35,475	2.9	Arabic English	Israel Syria Iraq Saudi Arabia
Kuwait	6,880	1.9	Arabic English	Iraq Saudi Arabia
Lebanon	4,015	2.7	Arabic French Armenian English	Syria Israel

Table 1-1- Desert countries of the world (cont).

COUNTRY	AREA SQ MILES	POPULATION (MILLIONS)	LANGUAGES	ADJACENT COUNTRIES
Libya	679,359	4.0	Arabic Italian English	Egypt Sudan Chad Niger Algeria Tunisia
Oman	82,031	1.3	Arabic English Baluchi Urdu	UAE Saudi Arabia Yemen
Qatar	4,247	.328	Arabic English	UAE Saudi Arabia Bahrain
Saudi Arabia	829,996	15.5	Arabic	Bahrain Iraq Jordan Yemen Oman UAE Qatar Kuwait
Syria	71,498	11.6	Arabic Kurdish	Lebanon Israel Turkey Iraq Jordan
Tunisia	63,170	7.7	Arabic French	Algeria Libya
United Arab Emirates	32,278	2.0	Arabic English Farsi Hindi Urdu	Qatar Saudi Arabia Oman

AUSTRALIA

DESERTS: GIBSON, GREAT SANDY, GREAT VICTORIA, SIMPSON

AUSTRALIA - GENERAL INFORMATION

Area	2,967,895 sq mi
Population Growth	1.17 %
Population Density	5/sq mi
Capital City	CANBERRA

AUSTRALIA - MAJOR CITIES

	Population	Latitude, Longitude
CANBERRA	274,000	35.18S, 149.08E
Sydney	3,441,000	33.53S, 151.10E
Melbourne	2,949,000	37.45S, 144.58E
Brisbane	1,169,000	27.30S, 153.00E
Perth	1,001,000	31.57S, 115.52E
Adelaide	987,000	34.56S, 138.36E
Newcastle	423,000	17.20S, 133.21E
Wollongong	238,000	17.13s, 137.57E
Gold Coast	208,000	27.59S, 153.22E
Hobart	180,000	42.54S, 147.18E
Geelong	150,000	38.10S, 144.26E

AUSTRALIA - LANGUAGES, ETHNIC GROUPS AND RELIGIONS

Languages	English	
	Aborigine	
Ethnic Groups	White	95%
	Asian	4%
	Aborigine and Other	1%
Religions	Anglican	26%
	Catholic	26%
	Other Christian	24%
	Other	24%

SOUTH AMERICA

CHILE

DESERTS: ATACAMA

CHILE - GENERAL INFORMATION

Area	756,945 sq km
Population Growth	1.49 %
Population Density	17 /sq km

CHILE - MAJOR CITIES

	Population	Latitude, Longitude
SANTIAGO	4,320,000	33.30s, 70.40W
Vina del Mar	320,000	33.02S, 71.35W
Valparaiso	273,000	33.05s, 71.40W
Talcahuano	225,000	36.40S, 73.10W
Conception	281,000	36.50S, 73.03W
Antofagasta	203,000	23.40S, 70.23W
Temuco	175,000	38.45S, 72.40W
Rancagua	157,000	34.10s, 70.45W
Talca	145,000	35.28S, 71.40W
Chillan	130,000	36.37S, 72.10W
Arica	158,000	18.30S, 70.20W

CHILE - LANGUAGES, ETHNIC GROUPS AND RELIGIONS

Languages	Spanish	
Ethnic Groups	European and Mestizo	95 %
	Native American	3%
	Other	2%
Religions	catholic	89%
	Protestant and Other	11 %

ASIA

CHINA

DESERTS: GOBI

CHINA - GENERAL INFORMATION

Area	9,600,000 Sq km
Population Growth	1.24 %
Population Density	113 /sq km
Capital City	BEIJING

CHINA - MAJOR CITIES

	Population	**Latitude, Longitude**
BEIJING	6,000,000	**39.55N, 116 26E**
Shanghai	7,000,000	**31.06N, 121 .22E**
Tianjin	5,400,000	**39.08N, 117 .12E**
Shenyang	4,200,000	**41.50N, 123 .26E**
Wuhan	3,400,000	**36.45N, 114 .15E**
Guangzhou	3,300,000	**23.08N, 113 .20E**
Chongqing	2,800,000	**29.30N, 106 .35E**
Harbin	2,630,000	**45.45N, 126 .41E**
Chengdu	2,580,000	**30.37N, 104 .06E**
Nanjing	2,250,000	32.03N, 118.47E

CHINA - LANGUAGES, ETHNIC GROUPS AND RELIGIONS

Languages	Mandarin	
	Cantonese	
	Shanghainese	
	Fuzhou	
	Minnan and Other	
Ethnic Groups	Han Chinese	**94%**
	Other	**6%**
Religions	Atheist and Eclectic	97 %
	Other	3%

MONGOLIA

DESERTS: GOBI

MONGOLIA - GENERAL INFORMATION

Area	1,564,619 sq km
Population Growth	2.76 %
Population Density	1 /sq km
Capital City	ULAN-BATOR

MONGOLIA - MAJOR CITIES

	Population	Latitude, Longitude
ULAN-BATOR	511,000	47.54N, 106.52E
Darhan	75,000	46.29N, 109.24E
Erdene	46,000	46.18N, 100.35E
Choybalsan	23,000	48.02N, 114.32E
Nalayh	15,000	47.40N, 107.12E
Ulangom	15,000	49.59N, 92.00E
Uliastay	14,000	47.42N, 96.52E

MONGOLIA - LANGUAGES, ETHNIC GROUPS AND RELIGIONS

Languages	Khalkha Mongol	
	Turkish	
	Russian	
	Chinese	
Ethnic Groups	Mongol	90%
	Kazakh	4%
	Chinese	2%
	Russian	2%
	Other	
Religions	Tibetan Buddhist	94%
	Muslim	4%
	Other	1%

PAKISTAN

DESERTS: THAR

PAKISTAN - GENERAL INFORMATION

Area	803,943 sq km
Population Growth	2.70 %
Population Density	134 /sq km
Capital City	ISLAMABAD

PAKISTAN - MAJOR CITIES

	Population	Latitude, Longitude
ISLAMABAD	270,000	33.40N, 73.08E
Karachi	5,400,000	24.51N, 67.02E
Lahore	3,132,000	31.34N, 74.22E
Faisalabad	888,000	31.25N, 73.09E
Rawalpindi	810,000	33.40N, 73.08E
Hyderabad	770,000	25.23N, 68.24E
Multan	750,000	30.10N, 71.36E
Gujranwala	700,000	32.06N, 74.11E
Peshawar	600,000	34.01N, 71.40E
Sialkot	310,000	32.29N, 74.35E
Sargodha	300,000	32.0IN, 72.40E

PAKISTAN - LANGUAGES, ETHNIC GROUPS AND RELIGIONS

Languages	Urdu	
	English	
	Punjabi	
	Sindhi	
	Pashtu and Other	
Ethnic Groups	Punjabi	66%
	Sindhi	13 %
	Pashtun	9%
	Baluchi and Other	12%
Religions	Sunni Muslim	77 %
	Shi'a Muslim	20 %
	Other	3%

AFGHANISTAN

DESERTS: KAR KUM

AFGHANISTAN - GENERAL INFORMATION

Area	249,999 sq mi
Population Growth	2.35 %
Population Density	58 /sq mi
Capital City	KABUL

AFGHANISTAN - MAJOR CITIES

	Population	Latitude, Longitude
KABUL	1,127,000	34.31N, 69.12E
Qandahar	195,000	31.35N, 65.45E
Herat	156,000	34.20N, 62.12E
Mazar-i-Sharif	115,000	36.42N, 67.06E
Jalalabad	111,000	34.26N, 70.28E
Qonduz	110,000	36,45N, 68.51E
Baghlan	109,000	36.11N, 68.44E
Maymana	40,000	35.53N, 64.38E
Pul-i-Khomri	35,000	35.55N, 68.45E
Ghazni	32,000	33.33N, 68.26E

AFGHANISTAN - LANGUAGES, ETHNIC GROUPS AND RELIGIONS

Languages	Pashtu	
	Dari	
	Turkish Languages	
	Baluchi	
	Pashai	
Ethnic Groups	Pashtun	50 %
	Tajik	25 %
	Uzbek	9%
	Hazara	9%
	Other	7 %
Religions	Sunni Muslim	74%
	Shi'a Muslim	15 %
	Other	11 %

SOUTHWEST ASIA

EGYPT

DESERTS: SAHARA

EGYPT - GENERAL INFORMATION

Area	1,001,449 sq km
Population Growth	2.65 %
Population Density	53 /sq km
Capital City	CAIRO

EGYPT - MAJOR CITIES

	Population	Latitude, Longitude
CAIRO	11,000,000	30.03N, 31.1

Alexandria	2,905,000	31.13N, 29.55E
Giza	1,640,000	30.01N, 31.10E
Shoubra El-Kheima	497,000	30.06N, 31.15E
El Mahalla El Koubra	355,000	30.59N, 31.10E
Tanta	344,000	30.48N, 31.00E
Port Said	364,000	31.17N, 32.18E
El-Mansoura	260,000	31,03N, 31.23E
Asyut	257,000	27.14N, 31.07E
Zagazig	256,000	30.35N, 31.30E
Suez	274,000	29.59N, 32.33E

EGYPT - LANGUAGES, ETHNIC GROUPS AND RELIGIONS

Languages	Arabic	
	English	
	French	
Ethnic Groups	Eastern Hamitic	90%
	Other	10%
Religions	Sunni Muslim	94%
	Other	6%

IRAN

DESERTS: KAVIR, LUT

IRAN - GENERAL INFORMATION

Area	1,648,000 sq km
Population Growth	3.23 %
Population Density	32 /sq km
Capital City	TEHRAN

IRAN - MAJOR CITIES

	Population	Latitude, Longitude
TEHRAN	6,028,000	35.40N, 51.2 6E
Mashad	1,120,000	36.41N, 52.39E
Isfahan	930,000	32.41N, 51.41E
Tabriz	855,000	38.05N, 46.18E
Shiraz	801,000	29.38N, 52.34E
Bakhtaran	535,000	31.28N, 54.54E
Karaj	527,000	35.48N, 50.58E
Ahwaz	471,000	31.17N, 48.43E
Qom	425,000	34.39N, 50.57E
Abadan	329,000	30.20N, 48.15E

IRAN - LANGUAGES, ETHNIC GROUPS AND RELIGIONS

Languages	Farsi
	Turkish

	Kurdish	
	Arabic	
	English	
Ethnic Groups	Persian	**63 %**
	Turkish	**18%**
	Kurdish	**3%**
	Arab and Other	16%
Religions	Shi'a Muslim	**93%**
	Sunni Muslim	**5%**
	Bahai and Other	**2%**

IRAQ

DESERTS: ARABIAN

IRAQ - GENERAL INFORMATION

Area	434,924 sq km
Population Growth	3.53 %
Population Density	40 /sq km
Capital City	BAGHDAD

IRAQ - MAJOR CITIES

	Population	**Latitude, Longitude**
BAGHDAD	3,250,000	**33,20N, 44.26E**
Basra	1,600,000	3030N,47.50E
Mosul	1,250,000	**36,21N, 43.08E**
Kirkuk	535,000	**35.28N, 44.26E**
An Najaf	141,000	**31.59N, 44.19E**
Erbil	99,000	**36.12N, 44.01E**

IRAQ - LANGUAGES, ETHNIC GROUPS AND RELIGIONS

Languages	Arabic	
	Kurdish	
	Assyrian	
	Armenian	
Ethnic Groups	Arab	**75 %**
	Kurdish	**17 %**
	Other	**8%**
Religions	Shi'a Muslim	**62%**
	Sunni Muslim	**35 %**
	Christian and Other	*3%*

LIBYA

DESERTS: SAHARA

LIBYA - GENERAL INFORMATION

Area	1,759,540 sq km
Population Growth	3.08 %
Population Density	2 /sq km
Capital City	TRIPOLI

LIBYA - MAJOR CITIES

	Population	Latitude, Longitude
TRIPOLI	610,000	32.49N, 13.07E
Misurata	320,000	32.07N, 20.04E
Benghazi	103,000	32.23N, 15.06E
Azzawiya	94,000	32.45N, 12.44E
Al-Beida	77,000	32.49N, 12.45E
Agedabia	51,000	30.48N, 20.15E
Dama	48,000	32.40N, 22.35E
Sebha	41,000	27.09N, 14.29E
Tubruq	35,000	32.05N, 20.30E
Al-Marj	30,000	32.25N, 20.30E
Zeleiten	26,000	32.28N, 14.34E

LIBYA - LANGUAGES, ETHNIC GROUPS AND RELIGIONS

Languages	Arabic	
	Italian	
	English	
Ethnic Groups	Arab Berber	97%
	Other	3%
Religions	Sunni Muslim	97%
	Other	3%

SYRIA

DESERTS: SYRIAN

SYRIA - GENERAL INFORMATION

Area	185,180 sq km
Population Growth	3.74 %
Population Density	62 /sq km
Capital City	DAMASCUS

SYRIA - MAJOR CITIES

	Population	Latitude, Longitude
DAMASCUS	1,250,000	**33.30N, 36.19E**
Aleppo	1,200,000	**36.14N, 37.10E**

Helms	430,000	34.44N, 37.17E
Latakaia	240,000	35.31N, 35.47E
Hama	200,000	35.05N, 36.40E
Deir ez-Zor	106,000	35.20N, 40.05E
Rakka	90,000	35.57N, 39.03E
Hasakeh	74,000	36.32N, 40.44E
Tartous	53,000	34.55N, 35.52E
Edleb	52,000	35.56N, 37.21E
Dera'a	50,000	32.37N, 36.06E

SYRIA - LANGUAGES, ETHNIC GROUPS AND RELIGIONS

Languages	Arabic	
	Kurdish	
	Armenian	
	French	
	English	
Ethnic Groups	Arab	90%
	Kurdish and Other	10%
Religions	Sunni Muslim	74 %
	Other Muslim	16 %
	Christian	10%

JORDAN

DESERTS: SYRIAN

JORDAN - GENERAL INFORMATION

Area	91,880 sq km
Population Growth	3.62 %
Population Density	31/sq km
Capital City	AMMAN

JORDAN - MAJOR CITIES

	Population	**Latitude, Longitude**
AMMAN	813,000	**31.57N, 35.56E**
Zarqa	277,000	**32.04N, 36.06E**
Irbid	141,000	**32.33N, 35.51E**
Ajhm	47,000	**32.20N, 35.45E**
Jarash	32,000	32.17N,35.45E
Madaba	26,000	31.44N, 35.48E

JORDAN - LANGUAGES, ETHNIC GROUPS AND RELIGIONS

Languages	Arabic	
	English	
Ethnic Groups	Arab	**98%**
	Circassian	**1%**

	Armenian	**1%**
Religions	Sunni Muslim	**95%**
	Christian	**5%**

DESERTS: ARABIAN

SAUDI ARABIA - GENERAL INFORMATION

Area	2,149,690 sq km
Population Growth	4.16 %
Population Density	7 /sq km
Capital City	RIYADH

SAUDI ARABIA - MAJOR CITIES

	Population	Latitude, Longitude
RIYADH	1,976,000	24.39N, 46.46E
Jeddah	1,084,000	21.30N, 39.10E
Mecca	810,000	21.26N, 39.49E
Taif	300,000	21.15N, 40.21E
Medina	250,000	24.30N, 39.35E
Dammam	175,000	26.25N, 50.06E
Hufuf	120,000	25.20N, 49.34E
Tabouk	80,000	28,22N, 36.32E
Buraidah	75,000	26.20N, 43.59E
Al-Mobarraz	60,000	25.26N, 49.37E
Kharmis-Mushait	54,000	18.19N, 42.45E

SAUDI ARABIA - LANGUAGES, ETHNIC GROUPS AND RELIGIONS

Languages	Arabic	
Ethnic Groups	Arab	90%
	Afro-Asian	**10%**
Religions	Muslim	**99%**
	Other	**1%**

SOUTH AFRICA

BOTSWANA

DESERTS: KALAHARI

BOTSWANA - GENERAL INFORMATION

Area	**231,804 sq mi**
Population 1988	1,190,000
Population Growth	3.49 %
Population Density	5 /sq mi
Capital City	GABORONE

BOTSWANA - MAJOR CITIES

	Population	Latitude, Longitude
GABORONE	95,000	24.45S, 25.55E
Francistown	34,000	21.07S, 27.32E
Selebi-Phikwe	33,000	21.58S, 27.48E
Serowe	25,000	22.25S, 26.44E
Mahalapye	22,000	23.05S, 26.51E
Molepolole	21,000	24.25S, 25.30E
Kanye	20,000	24.59S, 25.19E
Lobatse	20,000	25.11S, 25.40E
Mochudi	19,000	24.28S, 26.05E
Mauri	15,000	20.00S, 23.25E
Ramotswa	14,000	24.56S, 25.50E

BOTSWANA - LANGUAGES, ETHNIC GROUPS AND RELIGIONS

Languages	English	
	Setswana	
Ethnic Groups	Batswana	95 %
	Bushmen	4%
	European	1%
Religions	Indigenous Beliefs	50 %
	Christian	50 %

NAMIBIA

DESERTS: KALAHARA

NAMIBIA - GENERAL INFORMATION

Area	318,261 sq mi
Population Growth	5.30 %
Population Density	4 /sq mi
Capital City	WINDHOEK

NAMIBIA - MAJOR CITIES

	Population	Latitude, Longitude
WINDHOEK	100,000	**22.34S, 17.06E**
Tsumeb	13,000	**19.13S, 17.42E**
Keetmanshoop	11,000	**26.36S, 18.08E**
Otjiwarongo	9,000	**20.29S, 16.36E**
Luderitz	8,000	**26.38S, 15.10E**
Swakopmund	7,000	22.40S, 14.34E

NAMIBIA - LANGUAGES, ETHNIC GROUPS AND RELIGIONS

Languages	Afrikaans
	German
	English

	African Languages	
Ethnic Groups	Black	**86%**
	White	**7%**
	Mixed	**7%**
Religions	Christian	**60%**
	Indigenous Beliefs	**40%**

SOUTH AFRICA

DESERTS: KALAHARI

SOUTH AFRICA - GENERAL INFORMATION

Area	471,443 sq mi
Population Growth	2.20 %
Population Density	74 /sq mi
Capital City	PRETORIA

SOUTH AFRICA - MAJOR CITIES

	Population	**Latitude, Longitude**
PRETORIA	800,000	**25.45S, 28 12E**
Cape Town	860,000	**33 56S, 18 2 8E**
Durban	810,000	**29 53S, 31 00E**
Johannesburg	770,000	**26 10S, 28 02E**
Port Elizabeth	300,000	**33 58S, 25 3 6E**
Roodepoort	210,000	**26 10S, 27 5 3E**
Umhlazi	180,000	**29 53S, 31 00E**
Germiston	140,000	**26 15S, 28 10E**
Pietermaritzburg	130,000	**29 36S, 30 24E**
Boksburg	120,000	**26 13S, 28 15E**
Bloemfontein	110,000	29.07S,26.14E

SOUTH AFRICA LANGUAGES, ETHNIC GROUPS AND RELIGIONS

Language	Afrikaans	
	English	
	Zulu	
	Xhosa	
	Sotho and Other	
Ethnic Groups	African	**70%**
	White	18%
	Mixed	9%
	East Indian	3%
Religions	Christian	**60%**
	Hindu	2%
	Muslim and Other	**38%**

APPENDIX B

EMPLOYMENT OF AIRCRAFT IN DESERT OPERATIONS

The desert is probably the most severe of all environments in which aviation units must operate. Standard operating procedures for desert operations are different from areas having an abundance of contrasting terrain and substantial vegetation. This appendix describes some special considerations when employing aircraft in desert operations.

AIR OPERATIONS

Air combat operations that rely on heavy vegetation and varying contour terrain need to be flexible enough to incorporate different methods of camouflage and terrain flight techniques. The varying types of sand have a tremendous effect on operations-vast flat areas afford unlimited visibility, dunes are hard to distinguish at night, blowing sand impairs visibility and presents flight and maintenance problems, and surface composition affects the choice of landing zones, maintenance sites, FARPs, and operating bases. Additionally, low-hovering and taxiing aircraft generate blowing sand and dust that can cause aircrews to lose outside visual reference, and, if performed near other equipment, present additional maintenance problems for that equipment.

Air operations are not the only area affected by the desert environment. Aviation ground operations may require flexibility and modification to work around the heat and the effects of the terrain. Aircrews must resort to instrument flight during duststorms. Sand also causes excessive wearing, pitting, and eroding of aircraft components.

In certain areas, the desert, with its relatively level terrain and shallow compartments, contains few highly distinguishable terrain features to mask aviation forces. Formations of two or more aircraft can be seen 10 kilometers away because the dark airframes contrast against the desert sand. Aviation units normally deploy their aircraft along routes and may need to consider widely dispersed formations. Aviation forces can make maximum use of deception techniques during periods of limited visibility.

Air cavalry assets can conduct reconnaissance and security operations over great distances in the desert because of the lack of vegetation and relief. Even when they are sand painted, armor vehicles stand out starkly against the sand. When combined with traditional target acquisition principles, such as dust signature and movement, these factors make it easier to acquire and engage armored and mechanized forces well out of range of their main guns.

Aeroscouts flying nap of the earth (NOE) cannot necessarily find the enemy more easily than ground observers. Stationary targets are the most difficult to see as there is little to draw the observer's attention:

Therefore, aeroscouts must use caution to avoid blundering into enemy air defense weapons. The aircraft should land at a distance of 5-10 kilometers from the area of interest, and the observer should dismount and scan the area for suspected enemy. The observer must remain in contact with the pilot by using a portable radio. The process should be repeated at varying intervals until contact is made.

Attack helicopter battalions (ATKHBs) are a potent force in desert warfare. If they are employed quickly and violently, maximum results can be obtained both in offensive and defensive operations. They are best used where a quick concentration of combat power is needed. A desert environment presents excellent target acquisition and engagement possibilities. Attack assets must remain dispersed to provide security. Mission planning that incorporates flexibility is a key ingredient in the successful employment of ATKHBs.

Terrain flying and desert navigation require continuous concentration. Due to lack of terrain and poor reference points, the aviator may rely on dead reckoning, self-contained navigation equipment, and radio navigational aids. As light decreases, the ability to judge distances accurately is degraded and visual illusions become more common. Because of glare, haze, and frequently blowing sand, it may be difficult to detect changes in terrain and the horizon.

Attack helicopters should move from assembly areas to battle positions (or holding areas if necessary) over attack routes that will provide whatever cover and concealment and prominent terrain features necessary to assist in navigation and to decrease the possibility of detection. Attack helicopters may have multiple routes for ingress and egress. Route reconnaissance, premission planning, and prebriefs will maximize the benefits to route planning in desert operations.

The weather in desert regions can be extremely unpredictable. Sandstorms, accompanied by constantly fluctuating wind speeds, may reduce visibility from in excess of 50 kilometers to zero in less than five minutes. Pilots must be carefully briefed on prevailing weather conditions before takeoff. Warning of any expected variations in conditions must be transmitted immediately to all airborne aircraft.

THE PERFORMANCE OF HELICOPTERS IN HEAT

Aviation personnel must refer to the appropriate aircraft technical manuals to determine aircraft limitations and capabilities in the desert environment. Significant effects on the payload capabilities should be anticipated.

Commanders must develop realistic aircraft utilization procedures based on the environmental effects data provided by aviation staff personnel to obtain the fullest benefit from aviation assets.

Helicopters hovering close to the ground can cause the engine to ingest sand; can cause observation by the enemy due to the formation of dust clouds; or cause disorientation of the pilot due to blowing sand, particularly at night. Helicopters should not be moved under their own power while on the ground, but should be pushed or towed by men or vehicles. Maintenance should be restricted to the minimum time, and should take place on rock or on oiled or wet sand, if available. All apertures (Pitot tubes, for example) of aircraft should be covered when not in use (including helicopter windscreens).

Temperature and humidity have a direct impact on personnel and vehicle performance. Temperature and humidity affect air density. Air density decreases as temperature increases. High temperature and humidity reduce the efficiency of aircraft propulsion and aircraft lift capabilities. Although temperature and humidity may not have a direct effect on a particular operation, extremely high temperatures and humidity will reduce aircraft payloads.

APPENDIX C

TECHNIQUES FOR OPERATING EQUIPMENT IN THE DESERT

The effects of the desert environment on equipment were described in Chapter 1. This appendix describes techniques which, if used while operating equipment in the desert, can save both equipment and lives.

DRIVING

Drivers and track commanders should observe the guidelines in the following paragraphs while operating vehicles in desert areas.

Dusty Conditions

Wear goggles while driving open-hatched regardless of visibility. Clear-lens goggles should be worn at night unless night-vision goggles are used. Bandanas or surgical masks should be worn to avoid breathing heavy dust.

Vehicles in an extended convoy should maintain a dust distance of twice the normal interval, or as specified in the unit SOP to allow time for the dust to dissipate. When driving on extremely dusty roads or trails and if traffic conditions permit, a staggered column formation can be used with vehicles alternately driving on the left and right side of the road.

If the vehicles should become engulfed in dust, the convoy commander should consider adjusting the rate of march accordingly. Any commander of a vehicle engulfed in dust should alert the convoy commander by radio, move to the right side of the road, and stop or slow to allow the dust to dissipate. Extreme caution must be observed to ensure oncoming and following vehicles are not jeopardized. The lead vehicle must warn vehicles to return to column formation if encountering traffic.

Sandy Deserts

Sandy deserts may be relatively flat or interspersed with windblown dunes. When driving in sand, the following techniques should be applied:

- The best time to drive on sand is at night or early morning when the sand is damp and traction is better. However, this is not always the case especially with the newer type military tires with closer tread design. Damp sand packs between the tread in the grooves of these tires resulting in virtually no surface traction.

- Drivers of track vehicles must be wary of a lack of steering response, which indicates sand is building up between the rear sprockets and treads. If the buildup of sand is allowed to continue, it will force the tread off. "Shaking" the vehicle with the steering or backing up will remove the sand.
- Wheel vehicles may gain some traction by reducing the air pressure in the tires. However, prolonged driving on partially deflated bias ply tires will overheat the tires and break down the sidewalls. Vehicles equipped with radial tires or central tire inflation system (CTIS) are not affected by the lower tire pressure if the maximum speed listed in the operator's manual is not exceeded.
- Vehicle loads must be evenly distributed. Rear-wheel drive should be used where possible to prevent the front wheels from digging into the sand and becoming mired.
- Drivers must switch to all-wheel drive or change gears before a vehicle bogs down in the sand.
- Before entering the sand drivers should select a gear that will allow the vehicle to maintain as much torque as possible without causing the wheels to spin and to minimize changing gears.
- Large-wheeled vehicles, such as 5,000-gallon tankers, should have a designated "puller". The designated vehicle should be preconfigured to assist these vehicles when they become bogged down in loose sand.

Some sand areas will be covered by a surface crust. This is caused by chemicals in the ground cementing sand particles together, In many cases it will be possible to drive on top of this crust and minimize dust signature and the chance of bogging down. Consider the following techniques when driving on a crust:

- Use staggered columns to facilitate movement. As a general rule vehicles should not follow one behind the other.
- Ensure vehicles maintain a minimum speed (determined from experience) below which they will break through the crust.
- Avoid sharp turns and abrupt starts or stops that could cause a vehicle to break through the crust.
- Reconnoiter patches of the crust that are a different shade to ensure they are not softer than the surrounding crust.

Crossing Dunes

Crossing dunes requires careful reconnaissance. Normally, the upwind side of the dune will be covered with a crust and have a fairly gradual slope. The downwind side will be steeper and have no crust. Prior to crossing a dune, the driver should climb it on foot checking crust thickness, the angle at the crest to ensure the vehicle will not become bellied at the top, and the degree of slope and softness of the downwind side. If he is satisfied his vehicle can climb the dune, he should drive the vehicle straight up it at the best speed, crest it, and maintain a controlled descent on the other side.

Little hills may be formed by the wind blowing sand around small shrubs. Wheel vehicles should not attempt to move through areas where this has occurred without engineer assistance.

Cacti or thorn bushes will cause frequent tire punctures. Increase the number of tires carried in the unit's PLL when operating in areas covered with this type of vegetation.

Rocky Areas

Rock and boulder-strewn areas, including lava beds, may extend for many miles. Desert rocks, eroded and sharp-edged, vary in size and are so numerous that it is almost impossible to avoid any but the largest. The subsequent harsh jolting fatigues individuals and causes extreme wear on tracks, wheels, springs, and shock absorbers. Rocks and stones can become lodged between the tires on vehicles equipped with dusk that can cause severe damage to tires and brake components. Vehicles can follow one another in this type of terrain and it may be feasible to reconnoiter and mark a route. Drivers should achieve a "rolling" effect as they cross large rocks by braking as the vehicle's wheels ride over a rock so the axle settles relatively gently on the far side.

Salt Marshes

Salt marshes (called sebkha) are normally impassable, the worst type being those with a dry crust of silt on top. Marsh mud used on desert sand will, however, produce an excellent temporary road. Many desert areas have salt marshes either in the center of a drainage basin or near the seacoast. Old trails or paths may cross the marsh, which are visible during the dry season but not in the wet season. In the wet season standing water indicates trails due to the crust being too hard or too thick for the water to penetrate. However, such routes should not be tried by load-carrying vehicles without prior reconnaissance and marking.

RECOVERY

Track vehicle recovery methods are the same in the desert as in temperate climates. The techniques described in the following paragraphs will assist wheel vehicle recovery operations in sand crusts or salt marshes. To assist in recovery, wheel vehicles should carry the following items:

- Steel or aluminum channels, at least for the driving wheels. These are pierced to reduce weight and ribbed for strength. Pierced steel planking (PSP) or galvanized iron maybe used as a substitute.
- Sand mats made of canvas, preferably with lateral strips of metal to give strength and increase the traction of the wheels.
- Jacks and jack blocks.
- Tow rope(s).
- Shovels.

Once a vehicle becomes mired, excavate the ground under the vehicle in a gradual slope towards the direction of recovery to a point where no part of the underside is touching the ground. Channels or spurs and mats are laid under or against the wheels facing the direction of recovery. Tire pressure may be reduced to increase traction, but this also lowers the vehicle. It maybe necessary to lift the wheels with a jack if the vehicle is resting on its frame or axles.

When the vehicle begins to move, any faltering will cause it to sink again. Once out, the driver must maintain speed until the vehicle has reached the nearest hard area. At this point the tires are reinflated, the vehicle inspected for damage, and recovery equipment collected.

Vehicles equipped with winches can winch themselves out using ground anchors. The ground anchor may consist of a tarpaulin full of sand placed in a hole and the winch cable attached to it, or it may be one, or preferably two spare wheels well dug in.

A rubberized fabric balloon may be used on light vehicles to lift them free of broken crust. The balloon is placed under the vehicle and blown up with the vehicle exhaust.

If a lone vehicle breaks or bogs down in the desert, the crew must stay with it. A vehicle is much easier to find than a lone man.

MAINTENANCE

Equipment directly affected by heat, such as aircraft and radios, are equally affected by all deserts. However, power trains and suspension systems are affected in proportion to trafficability and soil texture. Most damage to equipment can be avoided by careful driving and by careful observation by vehicle commanders.

Track tension must be correct as constant driving on rocky plateau deserts will reduce the life of the track. Suspension units will require frequent replacement of torsion bars and suspension arms. To prevent damage to internal parts of the idler and suspension arms caused by the terrain, direct-support maintenance units must be provided with equipment capable of tapping and removing the sheared bolts.

To prevent problems that can result when desert vegetation clogs engine oil coolers and cylinder cooling pins, place a small-mesh wire screen over the top grille doors. It may still be necessary to remove packs about every 10 days to clean the engine cooling fins. The wire screening should be periodically checked, removed, and cleaned.

Maintenance personnel must inspect and adjust transmission bands frequently, especially on vehicles operating in hot, barren mountains. This will help reduce transmission oveheating.
Extra stocks of air-cooled generators are necessary because high-ambient temperatures limit their ability to maintain the proper operating temperature and contribute to premature failure.

NOTE: The M113A1 is especially susceptible to overheating problems in desert conditions. This includes the transmission and the solid-state voltage regulator, which is more prone to overheating

WHEELED VEHICLES

Wheeled vehicles are subject to brake system component failures and power steering leaks on rocky deserts. Vehicles equipped with manual transmissions are prone to clutch failure caused by drivers slipping the clutch. Vehicles with automatic transmissions tend to overheat; therefore, stop frequently to allow the transmission time to cool. The M54 5-ton truck is prone to air hydraulic cylinder failure and power steering seal leaks on rocky deserts. All vehicles of the 1/4- to 5-ton range are prone to clutch failure caused by drivers "riding" the clutch pedal. Tire consumption is very high, so all vehicles must carry one, or preferably two spare tires, and the unit's PLL of tires should be considerably increased. Approximately one vehicle in every three to four should carry slave cables to provide for battery failure.

Vehicles should be equipped with the following:

- Extra fan belts.
- Two spare tires.

Extra oil.

Extra radiator hoses.

Heavy duty tape.

Extra air and fuel filters.

- Jack stand support plate.
- Sand ladders (fabricated) and matting.
- Tow rope/cable.
- Extra water cans.
- Siphoning hose and funnel.
- Slave cables.

and early failure than the older mechanical type.

RADIOS

Radios, regardless of type, must be kept cool and clean. They must be in the shade whenever possible and should be located in a ventilated area (or even in an air-conditioned can). If water is available, wrap the radio in a damp towel, ensuring that the air vents are not blocked. Additional radios should be available in vital communications centers, such as tactical operations centers, to allow immediate replacement if the set in use shows signs of overheating.

It is essential that antennas be cut or adjusted to the length of the operating frequency. Directional antennas must be faced exactly in the required direction; approximate azimuth produced by guesswork will not do. A basic whip antenna relies on the capacitor effect between itself and the ground for efficient propagation. The electrical ground may be very poor, and the antenna performance alone may be degraded by as much as one-third if the surface soil lacks moisture, which is normally the case in the desert. If a ground-mounted antenna is not fitted with a counterpoise, the ground around it should be dampened using any fluid available. Vehicle-mounted antennas are more efficient if the mass of the vehicle is forward of the antennas and is oriented towards the distant station.

Desert operations require dispersion, yet the environment is likely to degrade the transmission range of radios, particularly VHF (FM) fitted with secure equipment. This degradation is most likely to recur in the hottest part of the day, approximately 1200 to 1700 hours.

If stations start to lose contact, especially if the hotter part of the day is approaching, alternative communication plans must be ready. Alternatives include the following:

- Using relay stations, including an airborne relay station (the aircraft must remain at least 4,000 meters behind the line of contact). Ground relay stations or RETRANS are also useful and should be planned in conjunction with the scheme of maneuver.
- Deploying any unemployed vehicle with a radio as a relay between stations.
- Using alternative radio links such as VHF multichannel telephones at higher levels, or HF (SSB) voice.
- Using wire. Normally wire will not be used as operations will be fluid, but it maybe of some value in some static defensive situations.
- Using a unit such as all or part of the task force scout platoon for messenger service, Although it is

undesirable to use such a unit in this manner, it may be necessary to maintain communications.

GENERAL TIPS

General tips for operating equipment in the desert include the following:

- Check track tension daily.
- Check drive belt adjustment frequently.
- Lubricate suspension items daily, and clean grease fittings.
- Reduce sand ingestion by stretching nylon stockings over air cleaners.
- Emphasize proper engine cooldown and shutdown procedures, especially diesels.
- Adjust battery specific gravity to the environment (refer to TMs).
- Set voltage regulators at lower end of specifications.
- Start up vehicles regularly to prevent discharge of batteries.
- Increase stocks of oils and lubricants.
- Use high-grade 20W-50 oil; it serves well under desert conditions.
- Compensate for increased pressure due to severe heat in closed pressurized systems.
- Check lubrication orders and TMs for the correct viscosity of lubricants for higher temperatures.
- Keep lubrication to the absolute minimum on exposed or semiexposed moving parts; this includes working parts of weapons.
- Erect screens against blowing sand in maintenance areas.
- Cover the gap between the fuel nozzle and the fuel tank filler neck opening during refueling operations.
- Protect exposed electrical cables and wires with electrical tape.
- Keep optics covered; clean them with a soft paintbrush or a low-pressure air system (this works well for weapons also).
- Clean sand and dirt from hulls of armored vehicles.
- Check tire pressures and fuel levels at the midpoint of the temperature range during the day.
- Ground all refueling equipment-STATIC ELECTRICITY KILLS.
- Replenish radiators with potable water whenever possible.
- Determine battery shortages early and requisition early.
- Drain fuel lines at night and in the morning due to condensation.
- Increase PLLs for the following parts due to high failure rates:

  - Tires.

  - All track components.

  -All suspension components for both wheel and track vehicles.

  - Brake shoes.

  - Bearings and bushings.

-Plastic and rubber parts, including seals.

- All filters.

- Generator components.

- Deploy with plastic bags to cover weapons and protect other equipment during maintenance or when not in use.
- Bring muzzle plugs.
- Prepare all vehicles for desert operations in accordance with the appropriate TMs.
- Issue small paintbrushes to all soldiers/marines for weapons cleaning and other equipment maintenance. The paintbrush is one of the more valuable tools available to the soldier/marine for maintenance.

APPENDIX D

EFFECTS OF THE ENVIRONMENT ON NBC WEAPONS

The primary purpose of using NBC weapons and smoke is to produce casualties, destroy or disable equipment, and generally disrupt operations. NBC weapons and smoke are employed in coordination with conventional weapons. FM 3-3, FM 3-4, FM 3-5, and FM 3-50 describe NBC defense and smoke operations. This appendix addresses the effects of a desert environment on NBC weapons.

WEATHER AND TERRAIN

Both desert weather and terrain affect the behavior of NBC and smoke weapons, and to some extent influence their tactical employment. The following paragraphs describe the effects of the weather and the terrain on the behavior of NBC weapons.

Weather

The effectiveness of NBC and smoke weapons is directly proportional to air stability. Air stability is a result of temperature variations at different levels of the air. The span of desert day and night temperatures causes extremes of air stability. At night and early morning the desert air is very stable. This may be the best time for NBC and smoke employment because of extensive downwind drift and area coverage. Desert air is very unstable during the late morning through afternoon. This may be the worst time for NBC and smoke employment because of quick and irregular dissipation. This may lead to ineffective target area coverage and possible danger to friendly troops. Temperature differences are determined by comparing the air temperature differences every 3.5 meters above the ground. Three types of temperature differences influence NBC and smoke operations:

- Unstable (lapse). Unstable conditions exist when air temperature decreases with altitude. In the desert, this mostly occurs between late morning and early evening.
- Neutral. Neutral conditions exist when air temperature does not change with altitude, In the desert, this mostly occurs during early morning and early evening.
- Stable (inversion). Stable (inversion) weather conditions exist when the air temperature increases with altitude. In the desert, this mostly occurs between late evening and early morning.

High desert temperatures in the middle of the day result in a decrease in air density, so nuclear blast waves move faster. Some chemical agents that come in contact with the skin are more effective during high temperatures due to perspiration. Desert sunlight is lethal to the effectiveness of most biological agents, but most are unaffected at night.

High winds are common in certain desert seasons and affect the dissemination of biological, chemical, smoke, and radioactive clouds. High winds may break up clouds in an odd fashion and force them in the wrong direction and possibly endanger friendly troops.

Terrain

Flat desert terrain lacking vegetation and under stable air conditions moves NBC and smoke effects evenly and steadily in all directions. Nuclear induced radiation is greater in sandy soil due to the high silica content. Desert soil below the surface crust is a fine powder, so the blast and suction effects of a nuclear burst will cause considerable dust clouds.

NUCLEAR WEAPONS IN THE DESERT

Many potential threat nations maintain nuclear weapons and additional nations continue to obtain nuclear capability. Nuclear weapons can be delivered by missiles or bombs and can be exploded in the air, on the ground, or below the ground. Depending upon the radiation dosage received, lethal effects could be felt immediately or delayed for days. Detonated nuclear weapons release energy that affects military troops and equipment in three forms: blast, nuclear radiation, and thermal radiation (heat and light). These effects are discussed in the following paragraphs.

Blast

Nuclear blasts, even from an air burst, raise considerable amounts of desert sand and dirt, which inhibits observation and maneuver for a long time. Lighter desert air density causes a drop in static overpressure, but a more rapidly expanding blast (shock) front. This increases the danger to aircraft and helicopters flying in the area. The radius of damage is normally smaller in desert climates so dug-in personnel and equipment are safer. Desert trafficability is degraded in the immediate area of the strike, especially for wheeled vehicles, due to the destruction of the sand crust.

Nuclear Radiation

Immediate nuclear radiation is a function of weapon yield and changes little with desert temperature. Residual radiation is high in the case of low air burst or ground burst weapons. Residual radiation in the desert is affected by the time of day and the wind. At night, with no wind, residual radiation may be evenly distributed around the point of burst. At night, with a steady wind, residual radiation may drift downwind for many miles. In the late afternoon, residual radiation may drift in a totally irregular pattern and direction due to desert air instability. Induced radiation is prevalent in desert sand due to the amount of silica in the soil. Constant radiation monitoring and reconnaissance are vital to protect troops, especially from contaminated water sources.

Thermal Radiation

The effective range of thermal radiation (heat and light) increases in the desert where there is little terrain masking. This increases the danger of troops receiving severe burns. Blinding light (dazzle), out to 50 kilometers, must be considered in the desert when determining the radius of warning for friendly troops.

BIOLOGICAL WEAPONS IN THE DESERT

The United States has renounced the use of biological agents, but many potential threat nations maintain biological weapons. Biological agents are living microorganisms that multiply inside the human body and cause disease. Biological agents may be disseminated as a liquid or vapor using rockets, bombs, or aerosol generators.

Depending upon the agent, they can produce lethal effects in 1-24 days from the time of exposure. High desert temperatures (120 degrees Fahrenheit and higher) and sunlight may destroy most biological agents. Cooler nighttime temperatures and the lack of sunlight provide biological agents a chance to enter the body. High desert winds will disperse biological agents more rapidly than low winds. Normally, the most effective wind speeds for effective target coverage are from 8-18 knots (14-32 kmph). Stable desert air conditions (night or early morning) provide the greatest agent concentration and area coverage. Unstable desert air conditions (late morning and afternoon) promote atmospheric mixing and lower agent concentration, reducing effective target coverage.

Biological weapons are best suited for strategic rather than tactical use in the desert. Test water and food sources frequently to ensure they are not contaminated. Sanitation, personal hygiene, and immunizations must be instituted to ensure individual protection.

The United States also renounces the use of toxins but some potential threat countries continue to develop toxins for possible military use. Toxins are extracted from natural biological sources and are not chemically produced. They can be disseminated as a liquid, vapor, or powder, and delivered by aerosol generators, artillery, rockets, or bombs. Depending upon the toxin, they can produce lethal effects in one minute to 12 hours from the time of exposure. In the desert, toxins can effectively be used as strategic or tactical weapons.

CHEMICAL WEAPONS IN THE DESERT

The United States has reserved the right to retaliate in kind with chemical agents against enemies who use them frost. There are six major types of chemical agents: nerve, blood, blister, choking, incapacitating, and tear. They are classified as persistent and nonpersistent. Chemical agents may be delivered by mines, rockets, artillery, or bombs. Depending upon the agent type, lethalities can occur between minutes to an hour from the time of attack.

High daytime temperatures of the desert increase incapacitating effects of liquid agents, which rely on skin penetration. Desert air instability (late morning and early afternoon) will usually cause quick, vertical and irregular dissipation of chemical agents. This reduces the lethal concentration and target area coverage. Desert air stability (early morning and evening) will most likely cause adequate agent dissipation and target area coverage. High desert temperatures also cause liquid agents to evaporate more quickly and decrease the time of hazard. If a liquid chemical agent soaks into the desert sand, it will increase the time of hazard. Figures D 1 and D-2 show the evaporation times of chemical agents (HD, GA, GF, VX) on sand at 104 and 68 degrees Fahrenheit for negligible risk levels. Negligible risk levels cause mild incapacitation among no more than 5 percent of unprotected soldiers/marines crossing a contaminated area.

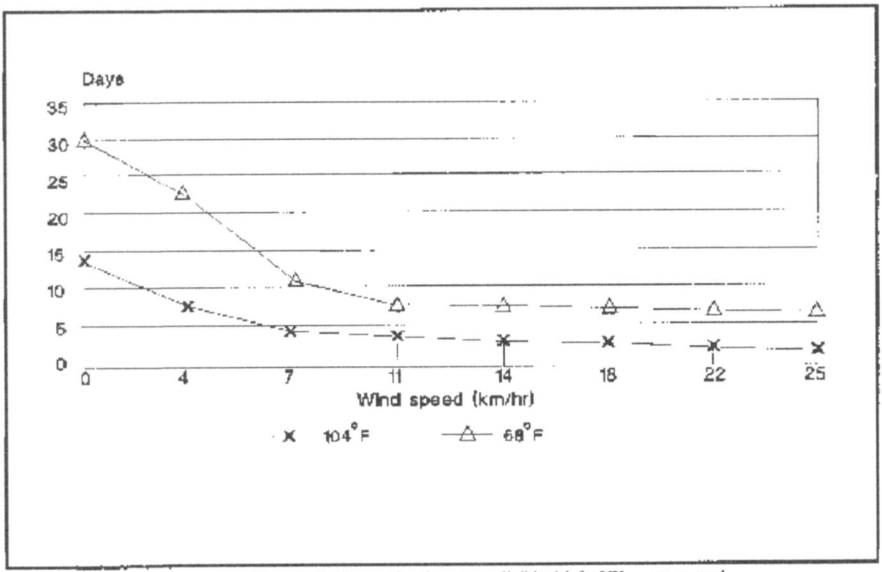

Figure D-1. Evaporation times to negligible risk for VX agents on sand.

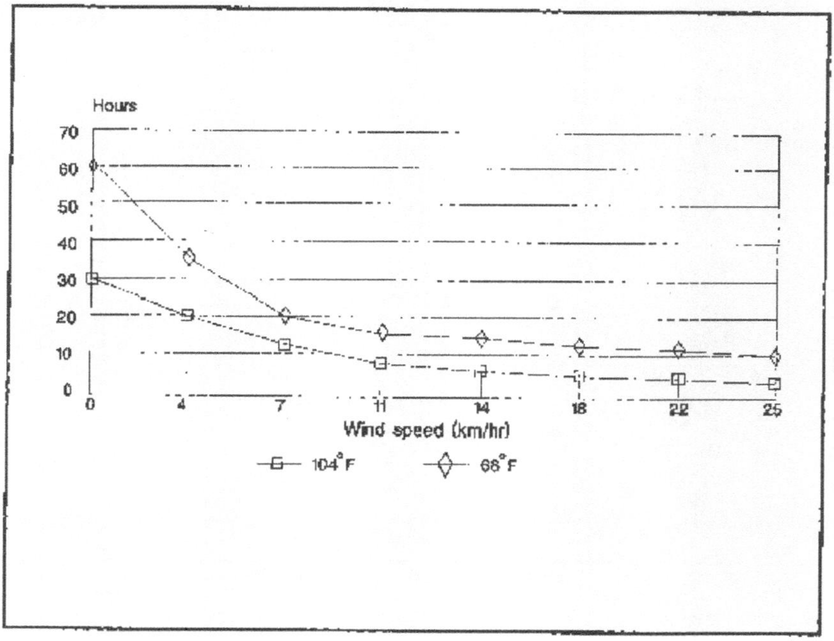

lot GF, HD, **and** GA agents.

Strong desert winds also increase the evaporation rate of liquid agents. They cause chemical clouds to drift downwind in an irregular direction and concentration. This tends to disperse target area coverage and lethal concentrations, and may endanger friendly troops.

SMOKE SYSTEMS IN THE DESERT

The lack of cover and concealment in flat desert terrain with little vegetation makes the use of smoke more vital to survival.

Desert winds, temperature, humidity, and terrain all affect smoke cloud behavior. The weather condition with the greatest impact on smoke operations is wind. Low desert winds allow smoke to remain on target areas for a longer period of time than high winds, In general, if the wind speed is less than 5 knots (9 kmph) or greater than 10 knots (18 kmph), smoke may not provide good target area coverage.

Temperature

Desert temperature differences have a direct relationship with making effective smoke. Three types of temperature differences influence smoke:

- Unstable (lapse). Lapse desert temperatures tend to break up and dissipate smoke, but are best for producing smoke curtains.
- Neutral. Neutral desert temperatures have limited vertical air currents and are good, but not the best, for producing smoke hazes and blankets.
- Stable. Stable (inversion) desert temperatures have no vertical air currents, and are the best for producing smoke hazes and blankets.

Humidity

Practically all smoke particles absorb moisture from the air. The lack of moisture in the desert air decreases the particle size and density of smoke making it less effective. Desert smoke streamers are shorter and less dense than smoke streamers in humid weather conditions. This increases the time and distance required to build adequate target area coverage.

TACTICS

The effects of NBC weapons and smoke on tactics in desert operations are discussed in the following paragraphs.

Nuclear

Nuclear weapons have a major bearing on tactics because of their ability to contaminate and shape desert terrain by making it impassable. Nuclear weapons can destroy troop and equipment concentrations and command and control centers and are usually considered weapons of mass destruction and combat multipliers; therefore, a smaller force with nuclear weapons may well defeat a much larger force without nuclear weapons.

The high maneuverability of tracked vehicles in the desert produces an endless number of avenues of approach. This creates difficulties for defense operations in the desert. Nuclear explosions could be used to shape desert terrain and canalize an enemy. This would provide profitable targets for other weapon systems. Equally, an attacking force could employ nuclear weapons in the desert to cut off enemy reinforcements, isolating them and making them vulnerable.

Biological Agents/Toxins

High desert temperatures and sunlight prevent the effective use of most biological agents during the daytime. An exception is spore-forming (anthrax) biological agents. During the day, troops crossing or occupying desert terrain face little danger from long-term biological contamination (except from spore-forming agents). But, because of favorable night desert conditions, an aerosol-delivered attack or a liquid biological attack would be effective. During night desert conditions, biological agents could effectively cover hundreds of square kilometers with a small amount of agent. This type of attack could be conducted covertly. These agents could contaminate vast areas of terrain and create mass casualties.

Toxins in the desert have nearly the same military use as chemical agents and are employed in the same manner. However, toxins are more deadly and require lower concentrations for lethalities. Toxins are difficult to detect with today's standard detection equipment.

Chemical Agents

The versatility of chemical agents gives commanders flexibility in desert operations. Commanders must consider how the employment of chemical agents affect offensive and defensive desert operations. Chemical agents can be used to create casualties, degrade performance, slow maneuver, restrict terrain, and disrupt logistical support. An assaulting force could use chemical agents to breach a defense or widen a gap. The best agent to use in this case would be a nonpersistent agent (one that is fast acting and leaves the target area quickly). High explosives could be mixed with the chemical attack to conceal the use of chemicals and complement their effects.

The threat of a chemical attack forces the use of protective masks and clothing. Heat, fatigue, and stress seriously affect the performance of troops. This is especially true with high desert temperatures. Well-trained soldiers/marines tolerate wearing protective gear better than those who are not as well trained. Troops in protective gear fire weapons less accurately, move more slowly, and must rest more often.

The actual or anticipated use of chemical weapons slows down a force and forces troops to take precautions. Desert heat, fatigue, and stress caused by wearing protective equipment slows down unit movement. Chemical agents can be used to create contaminated obstacles to desert maneuver. The best agent to use in this case would be a persistent chemical agent (one that remains on the target area for a time).

Chemical agents could be used to support offensive desert operations. Chemical agents could be used to protect flanks along an axis of advance to slow enemy counterattacks and to slow enemy fire and movement. In defensive desert operations, chemical agents could be used against second echelon forces to separate, slow down, and isolate them.

Chemical agents could be used to supplement conventional obstacles, or they could be used alone to restrict desert terrain. They may slow maneuver and channel attackers into engagement areas. A commander could contaminate a narrow desert mountain pass or bridge with a chemical agent and force the enemy to use an alternate route.

Logistical centers are lucrative targets for desert chemical attacks. Contaminating logistical supplies and equipment reduces the mobility of reinforcements and slows the delivery of supplies and equipment.

Smoke

Smoke is a combat multiplier that enhances the commander's ability to concentrate combat power at the critical time and place. Smoke is a far more significant battlefield factor than ever on flat desert terrain with little cover and concealment. Smoke can defeat enemy binoculars, weapon sites, and laser range finders in the desert. Smoke placed on the enemy at night interferes with enemy operations and observation by defeating enemy night sights and infrared sights.

In the desert offense, smoke can be used to deny the enemy information about the size, composition, and location of friendly maneuver forces. A smoke screen can be placed either to the front or to the flank. When the enemy cannot be screened effectively, obscuring smoke may be required. We must keep the enemy in doubt about the attacking unit's strength, position, activities, and movement. The longer the enemy is in doubt during an operation, the greater the chances are for mission success. Also, smoke can conceal desert breaching operations and river or gap crossings. It can also be used in deception operations.

In the desert defense, smoke is used to deny the enemy information about the size, location, and composition of friendly defensive positions. We must deny the enemy information by concealing the preparation and location of battle positions, artillery units, and reserves. Smoke can be used to support desert defensive positions by slowing enemy maneuver, disrupting command and control, isolating attacking echelons, silhouetting targets, and concealing obstacles.

PROTECTION

The commander of troops in desert operations must choose between the two following options when there is the threat of NBC warfare:

- Troops remain unprotected, with a high-chemical casualty risk, but a lower chance of heat fatigue.
- Troops are fully or partially protected, with a low-chemical casualty risk, but a higher chance of heat fatigue.

A decision on the level of protection is made according to the circumstances. If partial protection is ordered, the pace of physical work will be slower and proficiency reduced. The bulk of strenuous physical activity must be done at night or during the coolest part of the desert day.

In the desert, heat casualties (5 percent minimal) can be expected to occur in 30 minutes while performing heavy work in 90-degree-Fahrenheit temperatures when dressed in MOPP 4. Work/rest periods must be utilized to reduce the chances of heat fatigue. When protective clothing is worn, at least 10 degrees should be added to the WBGT index. Because of higher body temperatures, soldiers/marines in MOPP equipment perspire more than usual. Water must be consumed (2 quarts per hour) during continuous moderate work periods (and in MOPP equipment) when temperatures reach 80 degrees Fahrenheit and above to replace lost fluids or dehydration will follow.

DECONTAMINATION

The main problem of decontamination in the desert is lack of water, Although decontamination takes place as far forward as possible, the lack of water may burden the logistical system. Weathering may be a viable option for chemical contamination. The persistency of nerve agents (GA, GF, VX, GD) and mustard (HD) on chemical agent resistant coating (CARC) painted vehicles is between 4-24 hours at 104 degrees Fahrenheit. Sea water may be used as a substitute for fresh water during normal decontamination operations, but all equipment must eventually be flushed with fresh water to prevent corrosion.

Desert sand can be used for chemical decontamination, but it increases soldier/marine fatigue during its application. Sand removes most, but not all, liquid contamination, and saves valuable water supplies; however, the absorption capacity of desert sand is exhausted in 30-60 seconds after application. Remove the sand by sweeping or brushing the contaminated surface. Chemical agent detection should be conducted to ensure the agent is adequately removed.

APPENDIX E

DESERT CONCEALMENT AND CAMOUFLAGE

In the desert, camouflage problems are encountered that require imagination, ingenuity, and intelligence. The lack of natural overhead cover, the increased range of vision, and the bright tones of the desert terrain place emphasis upon siting, dispersion discipline, and the skillful employment of decoys to achieve deception and surprise. Total concealment is rarely achieved, yet proper camouflage measures can reduce the effectiveness of enemy observation, and consequently enemy operations.

Cover from enemy direct fire may be afforded by dunes, hills, and other irregularities in the desert terrain. Camouflage is an essential part of all operations in the desert and the importance of the concept must be impressed upon fresh units and individual replacements upon their arrival in theater. Poor camouflage may also compromise a high-level plan and lead to an operational failure. One poorly concealed vehicle can compromise an entire task force. Improvisation of available assets is just as important as being able to properly use camouflage systems. As previously described, deserts generally do not offer much natural concealment or means for camouflage; therefore, make maximum use of any artificial means available.

VEHICLES AND AIRCRAFT

Movement of vehicles produces dust, diesel plumes, and distinctive track marks. The slower the speed, the less dust that is produced; however, the need for speed must be balanced against the amount of dust that may be produced. Drivers must avoid harsh use of accelerators, the main cause of diesel plumes.

Shine from optical instruments (which should be kept shaded), and matte paint that has been polished by continual wear, or from tracks, particularly if rubber blocks have been removed, are difficult to camouflage during the desert day. See Figure E-l for shading optics. Running gear on tracks that has been polished by wear should be covered with burlap when stationary. Windscreens and windows should be removed or lowered to prevent reflection of the sun and heat. Vehicle silhouettes can be reduced in the forward areas by removing cabs and tops.

Figure E-1. shade optics to prevent shine

Disruptive pattern painting for vehicles and aircraft is described in FM 20-3. Local materials can also be used. The color and texture of the local terrain is best represented by placing dirt on vehicles and using a little water to make it stick.

The effects are increased by covering a vehicle with a wide-mesh net and using foliage brackets to attach local vegetation. Twine or wire may be used as an alternative to the mesh net, provided vegetation is available.

Some or all of the equipment listed in the following paragraphs should be available for every vehicle and aircraft, although aircraft will not necessarily be able to carry all of it.

The preferred net is the lightweight camouflage screen system (LCSS), desert version, which provides concealment against visual, near IR, and radar target acquisition/surveillance sensor devices. Additionally, the transparent version of the LCSS allows US units to camouflage radars (less CW type radars) without degrading operations. A desert camouflage net should be a complete cover, as it depends on its limitation of the ground surface, and both color and texture, for its effect. The alternatives to the LCSS in order of priority include the following:

- The specially produced desert-pattern net of the lightweight screen system.
- An open-weave cloth (colored as appropriate to the soil or "patched) stitched to an ordinary wide-mesh net and used with the string uppermost. This provides both color and texture and can be suitably garnished with radar-scattering plastic, such as that used in the lightweight screening system, and with any local vegetation.
- A cover of close-weave cloth, colored as appropriate.
- A standard net garnished solid, threaded in long straight strips that have been colored to harmonize with the terrain. The garnishing must be maintained.

The number of nets issued depends on the size of the equipment to be covered, but should be sufficient to allow a gradual slope of not more than 15 degrees from the top of the equipment to the earth. Each company-size unit should be equipped with a spray gun and various tints of paint to provide for temporary variations in net color to match the terrain.

When using nets for stationary equipment-

- Do not allow nets to touch sensitive items such as helicopter rotor heads and radio antennas which may cause a net to catch fire.
- Do not pull nets so tight that each supporting pole stands out.
- Ensure the net does not prevent the equipment from fulfilling its primary task. In some equipment, such as helicopters, a net must be easily removable to reduce reaction time.
- Avoid straight-edged patterns on the ground, which indicate something is there.
- Use burlap spray-painted in a nondescript desert color to cover all reflecting surfaces (excluding fire control optics) and shadow-producing areas under vehicle bodies, including tank suspensions. Aircraft equipped with windscreen covers will not require it.
- Cut desert scrub in the immediate area.
- Use poles, natural or man-made, to raise the nets from the equipment, thereby hiding its shape. They must be brought into the area of operations by the force and are extremely difficult to replace in the desert if lost or damaged.
- Make a "mushroom" out of thin iron tubing locally, It resembles an open umbrella without its cloth cover and with the end of the spokes joined together. Slotted into a socket that has been welded onto the top of a tank, self-propelled gun, or personnel carrier, it lifts the net above the vehicle, concealing its shape, increasing air circulation, and permitting the crew or team to use the top hatches.
- Hook and hold a camouflage net to the ground away from the vehicle by using wooden pegs or long steel pins, depending on soil consistency.
- Use mallets to drive pegs and pins into the ground.

After dismounting local security, camouflage is the first priority when a vehicle halts. Actions to be taken are-

- Site in vegetation or shadow, if available.
- Cover shiny surfaces and shadow areas with burlap screens.
- Drape the net.
- Add any available vegetation to the net.
- Blot out vehicle tracks for 50 meters behind vehicles.

Stationary aircraft take a relatively long time to conceal as they are fragile in comparison with other equipment, have a considerable heat signature, and must also be readily accessible for maintenance. The more they are concealed, the greater their response time is likely to be. Tactical flying is discussed in Appendix B, but take the following actions in sequence when approaching a landing site where aircraft will stay for some time:

- Ensure aircraft approach the site terrain-masked from enemy surveillance.

- Close down aircraft as soon as possible.
- Cover all reflective surfaces.
- Move aircraft into shadow if it can be towed or pushed.
- Shift the main rotor (depending on the type) until it is at a 45-degree angle with the fuselage and drape a net over the rotor and fuselage. The rotor must be picketed to the ground.
- Conceal the remainder of the aircraft.

POSITION SELECTION

Position selection is critical at all levels. One of the fundamentals of camouflage in any environment, but particularly the desert, is to fit into the existing ground pattern with a minimum of change to the terrain. A wadi bottom with vegetation or a pile of boulders that can be improved with grey burlap and chickenwire are good examples. Sites chosen must not be so obvious that they are virtually automatic targets for enemy suppressive frees, and antennas must be screened against the enemy, if possible.

Shadows, particularly in the morning and evening, identify objects; so equipment must be placed in total shadow (rarely found), or with its maximum vertical area facing the sun so that minimum shadow falls on the ground ("maximum vertical area" is the rear of a 5-ton truck with canopy, but the front of an M88, for example). See Figure E-2 for the effects of shadows. The shadow can be broken up, which is normally achieved by siting equipment next to scrub or broken surfaces, such as rocks. Equipment should not be sited broadside to the sun, and it is usually necessary to move as the sun moves. Digging in reduces the length of any shadow that is cast (on the principle that the lower the object, the shorter the shadow).

Figure E-2.Camoutlags: the effects of shadows.

Vehicles passing over pebbles or heavy ground surfaces press the pebbles or gravel into the soil, causing track marks to be prominent when viewed from the air. Avoid such areas if possible. Use existing trails and blend new trails into old ones whenever possible.

Soil texture suitable for digging must be a consideration when reconnoitering for battle positions. Holes must be covered to avoid shadows being cast. If vehicles will be in position for more than a day, trenches should be dug for them.

In forward areas, tactical operations centers are probably the most difficult positions to hide although their need for concealment is great. They require strict camouflage discipline. Vehicles and aircraft should not be allowed to approach closer than 300-400 meters. They must be dispersed and concealed so nets may have to be readily available for aircraft. Pay special attention to lights and noise at night.

Generators will have to be dug-in and allowed adequate air space for cooling. Radios and antenna systems must be remoted as far out as possible, and in different directions. Whenever possible, dig in the entire command post. Use engineer assets to build a berm around the perimeter and to help break up the silhouette and to enhance security. Other equipment should not be placed too close to minimize the possibility of the enemy's attention being attracted to the site.

Engineer activity often precedes operations, which makes it important that such work be concealed from enemy surveillance. The following guidelines should be used to conceal engineer activity:

- Employ the minimum number of equipment and personnel.
- Keep equipment well away from the site, and dispersed and concealed if not in use.
- Complete all possible preparations well away from the site.
- Follow the ground pattern, if possible.

Combat service and support assets must rely on concealment for most of their protection. The following guidelines will assist unit commanders in concealing trains while stationary or on the move:

- All vehicles of a given type should look alike. This will make it difficult for an enemy to pick out critical vehicles, such as water and fuel trucks, in a column. Canopies over fuel trucks disguise them and prevent radiant heat from striking the fuel containers.
- Vehicles should follow the tracks of the preceding vehicle if it is possible to do so without breaking through the crust, as this reduces the possibility of an enemy intelligence analyst to calculate how many vehicles have passed.
- Screen exhaust systems to reduce heat signature.
- Vehicles must never form a pattern, either when stationary or on the move.

SUPPLY POINTS

A supply point is likely to be in a location where its main threat of detection will be either by the eye or by photograph. Normally, greater emphasis can be placed on selecting supply positions from the point of view of concealment rather than for tactical efficiency, particularly in situations where air defense cover may be limited. The following guidelines should be used when setting up supply points:

- The location should be selected where trails already exist. Vehicles must use existing trails to the extent possible.

- Stocks should be irregularly spaced, both in length and depth, to the maximum extent possible so that there is no definite pattern.
- Stocks should be piled as low as possible and preferably dug-in. For example, a stack of gasoline cans should be only one can high.
- The shape of the area should not be square or rectangular, but should follow the local ground pattern.
- Stocks should be covered with sand, gravel, burlap, netting, or anything else that harmonizes with the local terrain, and the sides should be gradually sloped with soil filled to the top of the dump.
- The contents of each supply point should be mixed so that the destruction of one supply point will not cause an immediate shortage of a particular commodity.

APPENDIX F

OPERATIONS IN MOUNTAINS

Mountain operations are described in detail in FM 90-6. This appendix describes special conditions associated with operating in mountains such as those in the southern Sinai and on the shores of the Red Sea. It does not address tactics and techniques for mountain operations that are equally applicable to all mountains, except for the purpose of clarity.

TERRAIN

Mountains are high and rugged, with very steep slopes. Valleys running into a range become more and more narrow with the sides becoming gradually steeper. Valleys are usually the only routes that allow ground movement of men and equipment at any speed or in any quantity. Water is nonexistent on hilltops and unusual in valleys except during flash floods after rains. Lateral ground communications are limited unless the force is moving across the spines of mountain ranges. Navigation may be difficult, as maps are likely to be inaccurate.

PERSONNEL

Troops operating in mountainous country must be in peak physical condition. Regardless of their normal physical condition, personnel operating in mountainous areas require additional stamina and energy. They must also possess the ability to conduct sustained physical exertion and recover from it quickly.

Acclimatization is described in Chapter 1. Acclimatization to height, which varies much more among individuals than that for heat, must also be considered for operations in mountains. Lack of oxygen at high altitudes can cause unacclimatized troops to lose up to 50 percent of their normal physical efficiency when operating in altitudes over 6,000 feet. Mountain sickness may occur at altitudes over 7,800 feet and is usually characterized by severe headache, loss of appetite, nausea and dizziness, and may last from 5 to 7 days. Troops can acclimatize by appropriate staging techniques. It may take several weeks to become completely acclimatized, depending on altitude and the individual's personal physical reactions.

The risk of sunburn, particularly to the uncovered face, is greater in mountains than on the desert floor due to thinner atmosphere. Use antisunburn ointment and keep the face in shade around midday, using face nets or sweat rags. An individual camouflage net or scarf is particularly useful for this purpose. Recognition of heat illnesses in higher altitudes may not be as apparent as at lower altitudes because sweat evaporates very quickly. Measures to avoid dehydration and salt loss are extremely important. Daily temperature variations may be considerable making it necessary to ensure troops do not become chilled at night. Layering of clothing is essential. Troops who have been sweating heavily before the temperature starts to drop should take their wet shirts off and place them over relatively dry shirts and sweaters. Soldiers/marines should add layers of clothing as it gets colder and remove them as needed. This may have to be leader supervised and disciplined in the same manner as water consumption.

Requirements for hygiene areas important in mountainous areas as in the desert itself. Normal rocky ground will make it extremely difficult to dig any form of latrine so cover excrement with rocks in a specially marked area.

GENERAL CONSIDERATIONS

Infantry is the basic maneuver force in mountains. Mechanized infantry is confined to valleys and foothills (if these exist), but their ability to dismount and move on foot enables them to reach almost anywhere in the area. Airmobile infantry can also be extensively used. Consideration should be given to modifying the TOE of infantry units operating in barren mountains. A strong antitank platoon may not be necessary. However, the infantry requires extra radars and radios for the number of observation posts and separate positions that they may expect to occupy.

Mountains are not a good environment for tank and armored cavalry operations, because tanks and armored cavalry are unable to maximize their mobility, flexibility, and firepower.

Avenues of approach at ground level are few. Roads or trails are limited and require extensive engineer effort to maintain. Off-road trafficability varies from relatively easy to very difficult. Most movement and maneuver in this type of terrain is either by air or on foot. Unnecessary vertical foot or vehicle movement should be avoided. Rock slides and avalanches, although not as common as in high cold mountains, do exist and can restrict movement.

Air cavalry is the major reconnaissance means but they must guard against being ambushed by ground troops located at their own altitude or even higher. Security of units must include observation, especially at night, of all avenues of approach including those within the capabilities of skilled mountaineers.

It is relatively easier to conceal troops in barren mountains than on the desert floor due to rugged ground, deep shadows (especially at dawn and dusk), and the difficulties an observer encounters when establishing perspective. Carefully placed rocks can be used to hide equipment however, rocks can chip and splinter under small arms fire. The normal type camouflage net, which breaks up outline by shadow, maybe used rather than the overall cover normally used in the desert.

Helicopter units of all types can be used, although they maybe slightly inhibited by altitude and rugged terrain. Payloads and endurance are degraded due to density and attitude. Winds are turbulent with considerable fluctuations in air flow strength and direction, particularly on the lee side of mountains. These winds, combined with the terrain, produce extra strain on crews as they have little margin for error. Flight crews should receive training in these conditions before flying in operations under these conditions.

When using men on foot for navigation, use all available maps, the lensatic compass, and a pocket altimeter. The pocket altimeter is essentially a barometer, measuring height by means of varying air pressure. If a navigator can only establish his location in the horizontal plane by resection on one point, the altimeter tells him his height, and thus his exact position. The instrument must be reset at every known altitude as it is affected by fluctuations of air pressure. Air photographs can also be helpful if they are scaled and contoured.

Supply of water and ammunition and the evacuation of wounded, especially if helicopters cannot land, can complicate operations. Water and ammunition may have to be transported by unit or civilian porters using A-frames or other suitable devices, or even by animal transport such as camel or mule.

OFFENSIVE OPERATIONS

The objective in mountainous areas of operations is normally to dominate terrain from which the enemy can be pinned down and destroyed. Avenues of approach are normally few, with very limited lateral movement except by helicopter. Reconnaissance must be continuous using all available means, as enemy defensive positions will be difficult to find. Observation posts are emplaced on high ground, normally by helicopter.

When contact is made, airmobile infantry can be used to outflank and envelop the enemy while suppressive fires and close air support are placed on all suspected positions, especially on dominating ground. Engineers should be well forward to assist in clearing obstacles. If airmobile infantry is unable to outflank the enemy, it will be necessary to launch a deliberate attack.

Frontal attacks in daylight, even with considerable supporting frees, have a limited chance of success against a well-emplaced enemy. Flank attacks on foot take a lot of time. The best opportunity is at night or in very poor visibility, but progress of men on foot will be slow and objectives should be limited.

The force should make every effort to secure ground higher than enemy positions to allow the attack to be downhill. Mobile forces should select objectives to the enemy's rear to kill the enemy as they reposition or counterattack. Foot mobile forces must seek adequate terrain (restrictive) to equalize the enemy's mounted mobility advantage.

Air superiority is required to allow a continuous flow of supplies and combat support by helicopter. Friendly mobile units must concentrate to destroy enemy command and control, artillery, service support, and air defense assets. It may be possible to infiltrate to a position behind the enemy, preferably using the most difficult, and hence unlikely route. Although this is very slow, it normally has the advantage of surprise.

The importance of dominating terrain, together with the enemy's knowledge that troops on the objective will be physically tired and dehydrated, makes an immediate counterattack likely. Supporting weapons must be brought forward at once, preferably by helicopter, and casualties removed by the same method.

Airmobile and attack helicopter units are well suited for pursuit operations. They can be used to outflank retreating enemy, and set up positions overlooking likely withdrawal routes. Small engineer parties can be emplaced to block defiles and interdict trails. Close air support and field artillery are used to reinforce airmobile and attack helicopter units and to counter efforts by enemy engineers to create obstacles.

DEFENSIVE OPERATIONS

A defense from a series of strongpoints is normal in hot mountains due to the need to hold dominating terrain and restrictions on ground mobility. Due to the amount of rock in the soil, it takes more time to prepare positions and normally requires engineer support.

It is necessary to hold terrain dominating avenues of approach. Any terrain that dominates a friendly position must either be held, or denied to the enemy by fire. It may be necessary to stock several days' supplies, especially water, ammunition, and medical equipment in a position in case helicopters or supply vehicles are unable to reach it.

When a covering force is used, it is organized around cavalry reinforced with attack helicopters, supported by field and air defense artillery. Airmobile infantry operates on ridge lines. If the enemy closes on a battle position it is difficult to extract airmobile infantry, so sheltered landing sites nearby should be available. In any event, extractions must be covered by air or ground suppressive fires. Stay-behind observers should be used to call down field artillery fires on targets of opportunity or to report enemy activity. When tanks are a threat and terrain is suitable, the covering force is reinforced with tank-heavy units and antitank weapon systems.

Combat in the main battle area is usually a series of isolated actions fought from strongpoints on ridge lines and in valleys. Patrols are used extensively to harass the enemy and prevent infiltration; all possible routes must be covered. If the enemy attempts to outflank the friendly force, he must be blocked by attack helicopters, if available, or airmobile infantry.

Reserves should be kept centrally located and deployed by air to block or counterattack. If this is not possible, reserves may have to be split up and placed behind key terrain where they are available for immediate counterattack.

If retrograde operations are necessary, mountainous terrain is as good a place to conduct them as anywhere. More time is required to reconnoiter and prepare rearward positions, and they should be prestocked as much as possible. Unlike the desert floor where movement between positions is likely to cover relatively great distances, movement in these conditions is usually from ridge to ridge. Routes must be covered by flank guards, especially at defiles or other critical points, as the enemy will attempt to block them or cut off rear guards.

COMBAT SUPPORT

It may be difficult to find good gun positions at lower altitudes due to crest clearance problems-so high-angle fire is often used. The best weapons are light field artillery and mortars that are airmobile and can be manhandled so they can be positioned as high as possible.

Field artillery observation posts are emplaced on the highest ground available, although in low-cloud conditions it will be necessary to ensure that they are staggered in height. Predicted fire may be inaccurate due to rapidly changing weather conditions making observed fire a more sure method for achieving the desired results.

Like field artillery, there is limited use for self-propelled weapons in this environment, although some may be used in valleys. Airmobile towed weapons allow employment throughout the mountainous area of operations.

Major tasks for engineer, even in an airmobile force, are: construction, improvement, and route repair, and their denial to the enemy. Mining is important due to the limited number of routes. Lines of communication require constant drainage and possibly bridging to overcome the problem of flash flooding.

Because of the frequent interdictions of mountainous roadways, military police will experience multiple defile operations. Use temporary traffic signs to expedite traffic movement to the front. The number of stragglers may be expected to increase in this environment. Because of difficulty in resupply, the supply points for water, POL, food, and ammunition will become especially lucrative targets for enemy attack. Military police rear area security elements must develop plans for relief and for augmenting base defense forces.

COMBAT SERVICE SUPPORT

Air transportation is the best CSS means in mountain operations due to its mobility. It may be limited by the weather, enemy activity, or the scarcity of landing sites, so there should be alternative means available. Terrain permitting, wheel-vehicle transportation should be employed as far forward as possible, using high-mobility vehicles off main mutes. Beyond the limits of wheel transport the only alternatives to CSS transport are animals (which may need to be acclimatized) or porters.

The composition and employment of trains in mountain operations are described in FM 90-6. Brigade trains should locate near an airstrip that can handle USAF tactical airlift. They are an obvious target for enemy air attack or artillery, or raids by enemy deep patrols, so adequate air defense and a coordinated area defense plan are necessary. Guards must be placed on all dominating terrain around the area, equipped with ground surveillance radars and STANO devices, and patrols should be employed outside the perimeter.

Supply points may be set up in the brigade trains area to operate distribution points for Class I, III, and V supplies. However, where routes are limited it may be necessary to resupply totally by air from the DISCOM area.

The variations of supplies in demand in the desert are very much the same as for those in temperate climates and are described in Chapter 4. The differences are described below:

- Class I. Mess trucks are not practical in this terrain. Food is either eaten cold, or heated on can heaters. Each soldier/marine should carry a one-day supply of emergency rations to be used if the daily resupply does not arrive.
- Class II. There is a high demand for footwear. Combat boots may be expected to last approximately two weeks in the harsh rocky terrain.
- Class III. Individual vehicle consumption will be greater than normal. Aircraft fuel requirements are greater, but it should be possible for much of their refueling and servicing to take place well to the rear where resupply is relatively easy.
- Water continues to carry a high priority. Demand for water is approximately 9 quarts per day, per man, as a minimum, and sometimes considerably more. Troops should carry four canteens of water, and every effort should be made to prestock water in positions or along routes.

First aid at squad arid platoon level is very important as medics will not necessarily be able to reach individual isolated positions. It is easy to lose casualties in this terrain so a buddy system to keep watch on each individual should be a matter of SOP. Medical evacuation is most often by air. It is a comparatively long distance to the nearest helicopter landing site, so teams of stretcher bearers will be required.

APPENDIX G

WATER USAGE IN DESERT OPERATIONS

Combat operations in the desert pose a number of unique problems. Because there is so little water and because our troops and much of our equipment cannot survive without it, water is a critical item of supply in the desert. Forces trying to survive in the desert without adequate water supplies have always met with disaster. Finding and keeping water sources may be the most crucial issue in desert conflicts. At the very least, water sources will be critical.

PROPER USES OF WATER

Water must be used to support immediate and future missions. There will be times when there is enough water to fill all requirements, but there will also be times when command direction and considerable thought are necessary to decide how to make the best use of available water. The first priority must go to the survival of the force and accomplishment of the immediate mission, and second priority to the maintenance and sustainment of the force. A general priority for uses of water is--

- Personnel (drinking only).
- Medical treatment.
- Vehicle and equipment cooling systems.
- Personnel (uses other than drinking).
- Decontamination.
- Food preparation.
- Laundry.
- Construction.

How far the available water will stretch depends upon your evaluation of the local situation and mission; how you set priorities for water; and how careful you are in using your limited supply.

INTEGRATED BATTLEFIELD EFFECTS ON WATER USE

Water needs increase dramatically on a nuclear or chemical battlefield. Decontamination of men and equipment requires large quantities of water. Being "buttoned up" in vehicles or in NBC protective gear locks in heat and perspiration and makes men sweat more. This increases the need for drinking water to avoid dehydration and heat casualties. Replacement of fluids lost through sweat is a critical use of water and generally comes ahead of all other uses.

LEADER RESPONSIBILITIES

Your primary responsibility is to accomplish your mission. Water is essential to do that. You must estimate how much you need and when you need it, just as you do for any type of essential supply. When the supply is limited, you must adjust your plans. During the accomplishment of your mission you must monitor the supply and ensure that it is used according to your plan.

One of the biggest water-related problems you may face is that as troops recognize how valuable water is to their survival, they may hoard it and not drink enough to sustain the efforts you expect of them. Before periods of activity, have your troops drink as much as they can. During the activity, take positive steps to make soldiers/marines replace the water lost by sweating. Thirst is a poor indicator of the body's need for water and may be ignored during hard work or in the heat of battle. Squad and section leaders must make their men drink regularly.

EFFECTS OF HEAT AND LACK OF WATER

Objects absorb heat from the sun and the air. In the desert, heat from these two sources is extreme. The clear, low-humidity air lets most of the heat from the sun through. As the sunlight strikes an object, such as a soldier/marine or the ground, much of the heat is absorbed. The ground, heated by the sun, in turn heats the air, often to temperatures well over 100 degrees Fahrenheit. A man, or any other object in the sun, absorbs heat from both sources. A man in the shade only has to contend with the heat from the air. The third major source of heat is the body itself. Like an engine, the body generates excess heat as it functions. The more work performed, the more heat that is generated. There are four ways that heat leaves an object:

- Radiatio n heat radiates from an object to a cooler object through a medium.
- Conductio n heat flows from a hot object to a cooler object through direct contact.
- Convectio n heat flows from a hot object into a cooler surrounding medium like air.
- Evaporatio n heat is absorbed in changing a liquid (like water) into a vapor.

The heat losses from radiation and conduction are relatively small for people and equipment. If the air is much cooler than the surface of an object, such as a vehicle radiator or a person's skin, convection can remove significant amounts of heat. As the two temperatures get closer, this loss becomes smaller. If the air is hotter than the object, as is often the case in the desert, heat is gained from the air. In hot, dry desert climates, the primary method of losing excess heat from the body is the evaporation of sweat. The rate of sweating depends on the amount of excess heat the body needs to lose. Hard work in hot climates can result in 1-1/2 to 2-1/2 quarts of sweat lost per hour.

Supporting medical units measure the combined effects of the sun, air temperature, and humidity on dismounted troops in open terrain. This combined effect, measured as the Wet Bulb Globe Temperature (WBGT) or the Wet Globe Temperature (WGT), is translated by the supporting medical unit to a heat condition.

Men cannot be expected to routinely perform more than about 5 hours of heavy work per day in heat condition BLACK. However, using a 20-minute work/40-minute rest cycle, it will take 15 hours to do 5 hours of work. Similarly, in heat condition RED, a maximum of 6 hours of heavy work can be expected; in heat condition YELLOW, 7 hours; and in heat condition GREEN, 8 hours.

The heat condition experienced by individual soldiers/marines may differ from the general heat conditions in the area. The activity level in closed vehicles (such as tanks and armored personnel carriers) will probably be less than that of troops outside, but the air temperature may be 20 to 30 degrees Fahrenheit higher, thus increasing the soldier's heat gain from the air and, as a result, his sweat rate and need to drink water. As a general rule, increase the water intake to that for two or three heat conditions higher for troops in enclosed vehicles. On the other hand, troops performing light duty in the shade are not as severely stressed and will need less water and less rest time.

In practice, keeping track of the current heat condition and applying it to troops working in a wide variety of conditions will be difficult, if not impossible. This data gives you an understanding of how much water is required by troops when they work hard in the desert heat and to point out the limits of an individual's efforts in extreme heat. Work done in the heat of the day takes much longer and is more fatiguing than work done under relatively cooler conditions. Work schedules planned to take advantage of the cooler times of day-early morning, late evening, and night-not only increase productivity and reduce water use, but they are also easier on the men and better for their morale.

EFFECTS OF WATER LOSS

The body has a small reserve of water and can lose some without any effects. After a loss of about 2 quarts (which represents about 2.5 to 3.0 percent of body weight), effectiveness is impaired. Soldiers/marines may begin to stumble, become fatigued and unable to concentrate clearly, and develop headaches. Thirst will be present but not overpowering. So unless well trained, or reminded or goaded to drink, troops may not replace the water loss.

As dehydration continues, the effects will become more pronounced. The soldier/marine will become less and less effective and more likely to become a heat casualty. Some soldiers/marines will experience heat cramps, others will develop heat exhaustion or heatstroke. Heat cramps and heat exhaustion can be treated with good success and the soldier/marine returned to duty in a few days; however, without prompt medical attention, heatstroke can be fatal. Even if the man survives, he will probably not be returned to duty. In any case, a heat casualty is lost for some time. Preventing casualties is much easier than treating and replacing the casualties.

WATER SUPPLY PLANNING

Water planning is complicated because water is heavy (about 8.3 pounds per gallon) and may be considered perishable. Water stored in small containers gets hotter than water stored in large containers. As water gets hotter, it loses its disinfectant and becomes less desirable to drink. These facts make it difficult to carry an adequate supply of water, and frequent resupply is often required. The following questions must be answered when planning the unit's water supply:

- How much water is needed?
- Where is it needed?
- When is it needed?
- How will water get to the unit?
- How does water supply affect the mission?
- How does the mission affect water requirements?
- What measures need to be taken to ensure water is properly used?

There are several requirements for water. Some requirements, such as water for radiators, are reasonably constant. Some, such as water for food preparation or showers, are prescribed by the situation. Others, such as water for drinking and personal hygiene, depend on how the mission is accomplished. Planning water requirements for centralized service support functions (shower, laundry, medical treatment, maintenance, and construction) is the responsibility of the supporting organization. The largest and most critical planning factor is drinking water. The quantity required depends on the environment and the difficulty and intensity of individual activity.

When calculating water requirements for a whole day, you need to consider other requirements, such as shaving, brushing teeth, and helmet baths. On the average, these functions require almost 2 gallons of water per man per day. When B rations are issued, plan for 1 gallon of water per meal for the mess kit laundry, and 0.5 gallon per meal for food preparation and kitchen cleanup if the unit prepares its own B rations. The water used to heat individual combat rations can be reused for washing and shaving.

When calculating water requirements for individual details, plan to use 2 quarts of water per hour of hard work per man (including rest periods) during the heat of the day, or 1 quart of water per hour of hard work per man in the cooler parts of the day. These quantities are intended to satisfy requirements for drinking water as well as for the water that men will pour over their heads when they are hot. Experience with local conditions and the work performed may change these estimates. It is important to remember that water lost by sweating will be replaced sometime during the day, but men work best if the water is replaced as it is lost.

WATER REQUIREMENTS

Some examples of water planning calculations are provided in the following paragraphs.

Example 1. A 10-man squad performing heavy work that will take about 2 hours at a Continental United States (CONUS) base.

If the squad works in the late morning, with a 30-minute work, 30-minute rest cycle (heat condition RED), the work will take almost 4 hours. At 2 quarts of water per man per hour, the squad will require 20 gallons of water or 8 quarts (2 gallons) per man.

If the squad works at night or in the very early morning (heat condition GREEN or cooler), with a 50-minute work, 10-minute rest cycle, the work will take about 2 hours. At 1 quart of water per man per hour, the squad will require 5 gallons of water or 2 quarts (0.5 gallon) per man.

Example 2. A 40-man platoon doing a variety of work over an entire day.

The platoon requires 4 gallons of water per man per day for drinking, and 2.5 gallons per man per day for personal hygiene, or 6.5 gallons per man per day. For the whole platoon, 260 gallons (6.5 x 40) per day are required. (Meals not considered.)

Example 3. A 160-man company doing a variety of work and eating two B. ration meals and one MRE or MCI ration per day delivered from the battalion consolidated kitchen.

The company requires 4 gallons of water per man per day for drinking, 2.5 gallons per man per day for personal hygiene, and 1 gallon per meal for the mess kit laundry.

6.5 gallons (4 + 2.5) per man per day x 160 men	= 1,040 gallons
+ 1 gallon per B-ration meal x 320 rations*	= 320 gallons
	Total 1,360 gallons per day

*Not required if an expedient means can be used and utensils to avoid having to wash plates

NOTE: If company personnel operate their own field kitchen, 0.5 gallon per B-ration meal is required. In this example, an additional 160 gallons (0.5 x 320) would be needed.

Example 4. A 750-man battalion performing a variety of tasks, operating a battalion field kitchen, and feeding two B-ration meals and one MRE or MCI per day.

The battalion requires 6.5 gallons of water per man per day for drinking and personal hygiene, and 1.5 gallons per B-ration meal served.

6.5 gallons per man per day x 750	= 4,875 gallons
+ 1.5 gallons per B-ration meal x 1,500	= 2,250 gallons
	Total 7,125 gallons per day

The situation will usually dictate whether an element will pick up its own water or have it delivered to it. Normally, water is either produced at or delivered to a water supply point in the brigade rear. Forward units pick up their water at the water supply point and move it forward using organic transportation. Water distribution is normally planned and coordinated at company or battalion level.

Long distances and increased consumption compound water transportation problems in the desert. Several specialized pieces of equipment are available to overcome these problems. They have been designed and allocated to permit units to move and store the larger than normal volumes of water required in the desert. They are also capable of delivery by ground and air transportation, airdrop, or low altitude parachute extraction system (LAPES). Air transportation, while possible, is usually limited by aircraft availability and weight of water required. It is normally used only when ground transportation is not feasible. Water shortages can severely limit your unit's mission capabilities. If you cut water use to absolutely essential requirements, you can temporarily overcome limited water shortages, but severe shortages will limit your unit's mobility and capability. Water shortages may make some daylight operations or hard work infeasible or unsupportable. Each course of action must be analyzed with respect to water support requirements and the capacity of troops to sustain their efforts under severe heat conditions. Water supplies are equally important to the enemy. Taking and keeping water sources, or denying or destroying enemy water supplies can critically alter the options available to the enemy.

You and your unit can do several things to use water to your advantage. First, make sure that enough water is available for the most critical use-drinking. Furthermore, water must be available right where it is needed. Troops working in the desert should not have to walk more than a short distance to a water source. During rest periods some soldiers/marines would rather sit than walk to get water, especially if they are not thirsty. It must be easy for them to get water. Similarly, beverages must be readily available in dining facilities. There should be enough beverage dispensers that troops do not have to wait long in lines.

In addition to having water readily available, you must also make sure that soldiers/marines actually drink all the water they need. Since the signs of dehydration are not obvious until a person is close to heat injury, leaders at the squad and platoon level must keep track of water consumed and take the following measures:

- Have troops drink an extra quart of water before hard work. Storing water in your stomach gives you an extra quart of reserve capacity.
- Keep track of how much water each man drinks at the squad level; at the platoon level, monitor the use rate at each squad.
- Have troops take breaks as often as the heat condition requires, and during breaks remind or require the troops to drink.
- Make sure water is kept as cool as possible so it will be as palatable as possible to drink.
- Watch the troops for the first signs of heat stress and reduced effectiveness, such as stumbling and slurred speech.
- Have your soldiers/marines check their urine. A lack of the need to urinate and dark-colored urine are signs of dehydration.
- Use the buddy system within the squad to help ensure soldiers/marines are drinking enough.
- Make sure troops wear their uniform correctly. Shirts should be on with sleeves rolled down, scarf around neck, and hat on. The uniform should be worn as loosely as possible.

In order for water to be useful for its most critical purpose in the desert, it must be protected not only from enemy action but also from heat and contamination. The larger quantities of water required for drinking in the desert increase the importance of the quality of the water. Water can carry minerals, microbiological organisms, and toxic materials. The body can handle only so much of these contaminants before its natural defense mechanisms become swamped and health and effectiveness deteriorate.

When water has been treated and distributed to water points, it has already been checked for contamination by water purification unit operators and medics; however, these checks do not ensure the water will not become contaminated somewhere in the unit distribution system before a soldier/marine drinks it.

When water is purified and distributed, it is usually disinfected by adding chlorine to a level prescribed by the Command Surgeon. The chlorine not only kills the microbiological organisms presently in the water, but some also remains in the water to kill any bacteria that might get into the water later. Such contamination can be minimized by using common sense at the unit level. Store water only in clean containers intended for water, and do not let anything get in the water that you would not want to drink.

However, even if your unit is very careful, water may eventually become contaminated. As the water is handled or gets hot, the chlorine's disinfecting power disappears. Your company's field sanitation team measures chlorine levels in unit water containers. The team also has ampules of a chlorine compound that are used to replace chlorine in small containers. Each soldier/marine is issued a small bottle of iodine tablets to disinfect water in his canteen if he must take water from an expedient, untreated source.

Heat is another contaminant of water in the desert. When water is warmer than 75 or 80 degrees

Fahrenheit, it becomes difficult to drink. Bad tastes in water become more pronounced as water becomes warmer and people will not want to drink it. Small-unit leaders will have a very difficult task trying to get troops to drink all the water they need to replace losses if the water tastes bad. Water tastes best, and it is easier to drink large quantities of it, if it is between 50 and 70 degrees Fahrenheit. There are three ways to avoid the problems of drinking hot water:

- Drink it before it gets hot.
- Keep it in a cool place or in the shade.
- Cool it to a palatable temperature.

Drinking water before a period of heavy work or before leaving the unit area on a mission gets valuable water into the body before the water has a chance to heat up. It also provides an additional reserve of water that is easy to carry.

Water in containers absorbs heat from three sources-the air, the ground, and sunlight. There is no easy way to cut absorption from the air, but you can reduce heat absorbed from the ground and sun by keeping containers in as much shade as possible. There are certainly many ways of getting shade for water containers; with a little American ingenuity and available materials, it can be done.

Small, uninsulated water containers heat up more quickly than larger or insulated containers. Fortunately, the water in small containers can be used more quickly before it heats up. Water in uninsulated 5-gallon cans starting at a cool 60 degrees Fahrenheit can heat up to unpalatable temperatures in three to four hours on a hot (greater than 90 degrees Fahrenheit) day in the full sun, but it will take seven to eight hours if kept in the shade. Unshaded 55-gallon drums will heat to unpalatable temperatures in about one day of full sun, but will stay drinkable for two days if shaded. Insulated 400-gallon water trailers in or out of the sun will keep water cool for several days.

Water must be cool to start with if it is to be kept cool. One of the supplemental items available to company-size units is a small mobile water chiller. The chiller is designed to provide cool (about 60 degrees Fahrenheit), palatable water for company-size units. It can be used to cool and dispense water from any container into canteens, water cans, or any other container at a rate of at least 0.5 gallon per minute. It should be used in all cases where individuals must be encouraged to drink large quantities of water, and to cool water being taken with or to elements working away from the company area.

SURVIVAL TECHNIQUES

The water support logistics structure is designed to provide enough water to retain maximum force effectiveness, especially on a battlefield. It would be foolish to expect that the large quantities of water required for a combat force will always be available. Water is a critical supply in desert combat. Having less than full supply for all needs does not necessarily spell disaster for operations, but it may force changes in plans. There are several options available to you if you are forced to operate on reduced water supplies. The first part of this chapter outlines those options. The second part (described in the following paragraphs) describes actions and techniques for small groups or individuals to take when totally cut off from normal water supply.

The first and most obvious option is to eliminate, reduce, or postpone water uses not immediately required for survival or mission accomplishment. Showers, laundry, personal hygiene, and B-rations can usually be eliminated for several weeks without severe impact on troop health or combat effectiveness, although this tactic will eventually reduce troop health and morale. Also, construction requiring water can often be avoided.

The one water use that cannot be denied without significant risk is individual drinking. However, significant reductions in individual drinking requirements are possible by-

- Limiting all but absolutely essential work to the cooler parts of the day (early morning, late evening, and night).
- Keeping individuals in the shade as much as possible.
- Severely limiting all activity.

All of these actions reduce the effectiveness or capability of a force, but they help it to survive.

What can individuals or small groups do when they are totally cut off from normal water supply? If you are totally cut off from the normal water supply, the first question you must consider is whether you should try to walk to safety or stay put and hope for rescue. Walking requires 1 gallon of water for every 20 miles covered at night, and 2 gallons for every 20 miles covered during the day. Without any water and walking only at night, you may be able to cover 20 to 25 miles before you collapse. If your chance of being rescued is not increased by walking 20 miles, you may be better off staying put and surviving one to three days longer. If you do not know where you are going, do not try to walk with a limited supply of water.

If you decide to walk to safety, follow the following guidelines in addition to the general conservation practices listed in the next section:

- Take as much water as you have and can carry, and carry little or no food.
- Drink as much as you can comfortably hold before you set out.
- Walk only at night.

Whether you decide to walk or not, you should follow the principles listed below to conserve water in emergency situations:

- Avoid the sun. Stay in shade as much as possible. If you are walking, rest in shade during the day. This may require some ingenuity. You may want to use standard or improvised tents, lie under vehicles, or dig holes in the ground.
- Cease activity. Do not perform any work that you do not have to for survival.
- Remain clothed. It will reduce the water lost to evaporation.
- Shield yourself from excessive winds. Winds, though they feel good also increase the evaporation rate.
- Drink any potable water you have as you feel the urge. Saving it will not reduce your body's need for it or the rate at which you use it.
- Do not drink contaminated water from such sources as car radiators or urine. It will actually require more water to remove the waste material. Instead, in emergencies, use such water to soak your clothing as this reduces sweating.
- Do not eat unless you have plenty of water.

Do not count on finding water if you are stranded in the desert. Still, in certain cases, some water can be found. It does rain sometimes in the desert (although it may be 20 years between showers) and some

water will remain under the surface. Signs of possible water are green plants or dry lake beds. Sometimes water can be obtained in these places by digging down until the soil becomes moist and then waiting for water to seep into the hole. Desert trails lead from one water point to another, but they may be further apart than you can travel without water. Almost all soils contain some moisture.

APPENDIX H

FRATRICIDE REDUCTION

The problem of fratricide is as old as warfare itself. It is a complex problem that defies simple solutions. Fratricide is defined as "the employment of friendly weapons and munitions, with the intent to kill the enemy or destroy his equipment or facilities that results in unforeseen and unintentional death or injury to friendly personnel." This is obviously a broad definition. This appendix focuses on actions leaders can take with current resources to reduce the risk of fratricide.

MAGNITUDE OF THE PROBLEM

The modern battlefield is more lethal than any in history. The pace of operations is rapid, and the non-linear nature of the battlefield creates command and control challenges for all unit leaders.

Our ability to acquire targets using thermal imagery exceeds our ability to accurately identify targets as friend or foe. The accuracy and lethality of modem weapons make it possible to engage and destroy targets at these extended acquisition ranges.

Added to this is the problem of battlefield obscuration. Rain, dust, fog, smoke, and snow degrade the ability to identify targets by reducing the intensity and clarity of thermal images. The effects of battlefield obscuration must be considered when thermal identification is relied upon.

On the battlefield, positive visual identification cannot be the sole engagement criteria at ranges beyond 1,000 meters. Situational awareness is key and must be maintained throughout an operation.

The following are recommended actions to take at crew and leader level in the event the crew are victims of friendly fires:

- React to contact until you recognize friendly fire.
- Cease fire.
- Report on the next higher unit net¬

    - That you are receiving friendly fire.

    -The location and direction of the firing vehicle.
- Provide a visual recognition signal to cease fire.
- Protect troops, request medical assistance as needed.
- Do not return fire when you positively identify the firing unit as friendly.

The following are recommended actions to take at crew and leader level when the crew are engaging friendly forces:

- Cease fire.
- Report on next higher net—
  - The engaged friendly force (if unknown, report number and type of vehicles).
  - The location.
  - The direction and distance to victim.
  - The type of fire.
  - The target effects.

The following are recommended actions to take at crew and leader level in the event the crew observes a friendly fire incident:

- Seek cover and protect self.
- Report on next higher net—
  - The friendly force engaged.
  - The location of the incident.
  - The direction and distance to victim/firer.
  - The type of fire.
  - The target effects.
- Provide a visual friendly recognition signal.
- Provide assistance (when safe to do so) as needed.

Leader actions should focus on identifying and stopping the friendly fire incident and establishing controls to prevent its recurrence. Some recommended actions for identifying and stopping friendly fire incidents are-

- Find and stop firing.
- Conduct in-stride risk assessment.
- Implement controls to preclude recurrence.

PREVENTIVE MEASURES

Reduction of fratricide risk begins with the planning phase of an operation and continues through the execution of the operation. The following are considerations for identifying fratricide risks in the planning, preparation, and execution phases of a given operation:

- Planning phase. A good plan that is well understood helps to minimize fratricide risk. The following considerations help indicate the potential for fratricide in a given operation:
  - The clarity of the enemy situation.
  - The clarity of the friendly situation.

- The clarity of the commander's intent.
- The complexity of the operation.
- The planning time available to all levels.

• Preparation phase. The following additional fratricide risks may become evident during rehearsals:

- Number and type of rehearsals.
- Training and proficiency levels of unit/individuals.
- The habitual relationships between units conducting the operation.
- The endurance of the troops conducting the operation.

• Execution phase. During execution, in-stride risk assessment and reaction are necessary to overcome unforeseen Fratricide risk situations. The following are factors to consider when assessing fratricide risks:

- Intervisibility between adjacent units.
- Amount of battlefield obscuration.
- Ability or inability to positively identify targets.
- Equipment similarities and dissimilarities between enemy and friendly vehicles.
- Vehicle density on the battlefield.
- The tempo of the battle.

Graphics are a basic tool that commanders at all levels use to clarify their intent, add precision to their concept, and communicate their plan to subordinates. As such, graphics can be a very useful tool in reducing the risk of fratricide. Commanders at all levels must understand the definitions and purpose of operational graphics and the techniques of their employment. See FM 100-5-1 for the definitions of each type of graphic control measure.

Briefbacks and rehearsals are primary tools in identifying and reducing fratricide risk. The following are some considerations on briefbacks and rehearsals to aid in reducing fratricide:

• Briefbacks ensure subordinates understand their commander's intent. They often highlight areas of confusion, complexity, or planning errors.
• The type of rehearsal conducted impacts on the risks identified.
• Rehearsals should extend to all levels of command and involve all key players.
• Use briefbacks or rehearsals to ensure subordinates know where fratricide risks exist, and what to do to reduce or eliminate the risk.

Maintaining situational awareness at all levels is key to fratricide reduction. Units must develop techniques to gain and maintain situational awareness in SOPs. Techniques could include-

• Eavesdropping on next higher net.
• Cross talk on radio between units.
• Accurate position reporting and navigation.

- Training and use/exchange of LOS.

Risk assessment must be conducted at all levels during the planning, preparation, and execution phases of all operations. Identification of fratricide risk factors is conducted at every level and the results should be clearly communicated up and down the chain of command.

Figure H-1 provides a worksheet for considering fratricide risk in the context of mission requirements. The worksheet lists 26 mission-accomplishment factors that affect the risk of fratricide. Assess the potential risk in each area as low, medium, or high, and assign a point value to each (one point for low risk, two for medium risk, three for high risk). Add the point values for the overall fratricide assessment score. Use the resulting score only as a guide, however. Your final assessment must be based both on observable risk factors like those on the worksheet and on your "feel" for the intangible factors affecting the operation. Note that descriptive terms are listed only in the low- and high-risk columns of the worksheet. Your assessment of each factor will determine whether the risk matches one of these extremes or lies somewhere between them as a medium risk.

The following fratricide reduction measures are provided as reminders for prudent/appropriate actions to reduce fratricide risk. They are not directive in nature, nor intended to restrict initiative. Apply the following measures to METT-T situations as appropriate:

- Identify and assess potential fratricide risk in estimate of the situation. Express this risk in the OPORD or FRAGO.
- Maintain situational awareness-current intelligence; unit locations/dispositions; denial areas (minefields/FASCAM); contaminated areas (e.g., ICM and NBC); SITREPs; and METT-T.
- Ensure positive target identification. Review vehicle/weapons ID cards. Know at what ranges and under what conditions positive ID of friendly vehicles/weapons is possible.
- Establish a command climate that stresses fratricide prevention. Enforce fratricide prevention measures; use doctrinally sound tactics, techniques, and procedures to ensure constant supervision of execution of orders and performance to standards.
- Recognize the signs of battlefield stress. Take quick effective action to deal with it to maintain unit cohesion.
- Conduct individual and collective (unit) fratricide awareness training; target identification/recognition training; fire discipline; and leader training.
- Develop a simple decisive plan.
- Give complete and concise mission orders.
- Use SOPS that are consistent with doctrine to simplify mission orders. Periodically review and change SOPS as needed.
- Strive for maximum planning time for you and your subordinates.
- Use common language/vocabulary and doctrinally correct standard terminology and control measures, such as, fire support coordination line (FSCL), zone of engagement, restrictive fire line (RFL), and others.
- Ensure thorough coordination is performed.
- Plan for and establish good communications.
- Plan for collocation of command posts, as appropriate to the mission (e.g., passage of lines, and so

forth).

- Establis h and designat e liaison officers (LO) as appropriate.
- Mak e sure rules of engagement (ROE) are clear.
- Conside r the effect of fratricide on key elements of terrain analysis (observation and fields of fires, cover and concealment, obstacles and movement, key terrain, and avenues of approach).
- Conduc t rehearsals whenever the situation allows time to do so.
- B e in the right place at the right time. Use position location/navigation (GPS) devices; know your location and the locations of adjacent units (left, right, leading and follow on); and synchronize tactical movement.
- Includ e fratricide incidents in after-action reviews (AAR).

FACTORS	LOW (1)	MEDIUM (2)	HIGH (3)
1 UNDERSTAND PLAN			
Commander's Intent	Clear		Foggy
– Complexity	Simple		Complex
– Enemy Situation	Known		Unknown
Friendly Situation	Clear		Unclear
– ROE	Clear		Unclear
2 ENVIRONMENT			
– Intervisibility	Favorable		Unfavorable
– Obscuration	Clear		Obscured
– Battle tempo	Slow		Fast
– Positive target ID	100%		0%
3 CONTROL MEASURES			
– Command relationships	Organic		Joint/Combined
– Audio	Loud/Clear		Jammed
– Visual	Well Seen		Obscured
– Graphic	Standard		Not understood
– SOPs	Standard		Not used
– LOs	Proficient		Untrained
Location/Navigation	Sure		Unsure
4 EQUIPMENT (Compared to US)			
– Friendly	Similar		Different
– Enemy	Different		Similar
5. TRAINING			
Individual proficiency	MOS Qual		Untrained
Unit proficiency	Trained		Untrained
– Rehearsal	Multiple		None

– [illegible]	[illegible]		[illegible]
– Habitual relationship	Yes		No
– Endurance	Alert		Fatigued
6 PLANNING TIME (1/3 -2/3 Rule)			
– Higher HQ	Adequate		Inadequate
– Own HQ	Adequate		Inadequate
– Lower HQ	Adequate		Inadequate
OVERALL FRATRICIDE ASSESSMENT	LOW 26–46%*	MEDIUM 42–62%*	HIGH 58–78%*

* Commander may use numbers as the situation dictates. Numbers alone may not give accurate fratricide risk.

Figure H-1 Fratricide risk assessment worksheet.

FRATRICIDE RISK CONSIDERATIONS

This format, which parallels the five-paragraph OPORD, contains key factors and considerations in fratricide reduction, This is not a change to the OPORD format; rather, it should be used during OPORD development to ensure fratricide reduction measures are included in the order. It is not a strict guide. The factors and considerations are listed where they would likely appear in the OPORD, but they may warrant evaluation during preparation of other paragraphs.

1. Situation.

a. Enemy forces.

(1) Are there similarities between enemy and friendly equipment and uniforms that could lead to fratricide?

(2) What languages do enemy forces speak? Could these contribute to fratricide risk?

(3) What are the enemy's deception capabilities and its past record of deception activites?

(4) Do you know the locations of enemy forces?

b. Friendly forces.

(1) Among the allied forces, are there differences (or similarities with enemy forces) in language, uniform, and equipment that could increase fratricide risk during combined operations?

(2) Could differences in equipment and uniforms among US armed forces increase fratricide risk during joint operations?

(3) What differences in equipment and uniforms can be stressed to help prevent fratricide?

(4) What is the friendly deception plan?

(5) What are the locations of your unit and adjacent units (left, right, leading, follow-on)?

(6) What are the locations of neutrals and noncombatants?

c. Own forces.

(1) What is the status of training activities? What are the levels of individual, crew, and unit proficiency?

(2) Will fatigue be a factor for friendly forces during the operation? Has an effective sleep plan been developed?

(3) Are friendly forces acclimatized to the area of operations?

(4) What is the age (new, old, or mix) and condition of equipment in friendly units? What is the status of new equipment training (NET)?

(5) What are the expected MOPP requirements for the operation?

d. Attachments and detachments.

(1) Do attached elements know the above information regarding enemy and friendly forces?

(2) Are detached elements supplied the above information by their gaining units?

e. Weather.

(1) What are the expected visibility conditions (light data and precipitation) for the operation?

(2) What effect will heat and cold have on troops, weapons, and equipment?

f. Terrain.

(1) Do you know the topography and vegetation (such as urban, mountains, hilly, rolling, flat, desert, swamp/marsh, prairie/steppe, jungle, dense forest, open woods) of the expected area of operations?

(2) Have you evaluated the terrain using the factors of OCOKA?

2. Mission. Is the mission, as well as all associated tasks and purposes, clearly understood?

3. Execution.

a. Task organization.

(1) Has the unit worked under this task organization before?

(2) Are SOPS compatible with the task organization (especially with attached units)?

(3) Are special markings or signals (for example, cats' eyes, chemlites, or panels) needed for positive identification of uniforms and equipment?

(4) What special weapons and/or equipment are to be used? Do they look or sound like enemy weapons and/or equipment?

b. Concept of the operation.

(1) Maneuver. Are main and supporting efforts identified to ensure awareness of fratricide risks and prevention?

(2) Fires (direct and indirect).

(a) Are priorities of fires identified?

(b) Have target lists been developed?

(c) Has the fire execution matrix/overlay been developed?

(d) Have locations of denial areas (minefields/FASCAM) and contaminated areas (ICM, NBC) been identified?

(e) Are the locations of all supporting fires targets identified in the OPORD/OPLAN overlays?

(f) Are aviation and CAS targets clearly identified?

(g) Has the direct-fire plan been developed?

(h) Have final protective fires (FPF) been designated?

(i) Have you identified and verified sector limits?

(3) Engineer tasks.

(a) Are friendly minefield, including FASCAM and ICM dud-contaminated areas, known?

(b) Are obstacles identified, along with the approximate time needed for reduction/breaching of each?

(4) Tasks to each subordinate unit. Are friendly forces identified, as appropriate, for each subordinate maneuver element?

(5) Tasks to CS/CSS units. Have locations of friendly forces been reported to CS/CSS units?

(6) Coordinating instructions.

(a) Will a rehearsal be conducted? Is it necessary? Are direct and indirect fires included?

(b) Is a briefback necessary?

(c) Are appropriate control measures clearly explained and illustrated in the OPORD and overlays? Have they been disseminated to everyone who has a need to know? What is the plan for using these control measures to synchronize the battle and prevent fratricide?

(d) Have target/vehicle identification drills been practiced?

(e) Do subordinate units know the immediate action, drill, or signal for "cease fire" or "I am friendly" if they come under unknown or friendly fire? Is there a backup action?

(f) Is guidance in handling dud munitions (e.g., ICMs and CBUs) included?

4. Service Support.

a. Are trains locations and identification markings known by everyone?

b. Do medical and maintenance personnel know the routes between train units?

5. Command and Signal.

a. Command.

(1) What is the location of the commander and key staff?

(2) What is the chain of command?

b. Signal.

(1) Do instructions include signals for special and emergency events?

(2) Do instructions include how to identify friendly forces to aircraft?

(3) Do instructions include backup code words and visual signals for all special and emergency events?

(4) Are signal operation instructions (SOI) distributed to all units with a need to know (e.g., higher, lower, adjacent, leading, follow-on).

A2C2	Army airspace command and control
AA	assembly area
AAA	antiaircraft artillery
AAFAD	all army for air defense
AFCE	Allied Air Farces Central Europe
A AG	army artillery group
AAR	after-action review
AC	analysis console
ACA	airspace coordination areas
ACC	air component commander
ACP	Allied communication Publication
ACR	armored cavalry regiment
ACT	air cavalry troop
AD	armored division
ADA	air defense artillery
ADCOORD	air defense coordinator
ADW	air defense warming
AH	Air Force
AFAC	airborne forward air controller
AFCENT	Allied Forres Central Europe
ARO	aerial fire support officer
AFV	armored fighting vehicle
AG	adjutant general
AGOS	air-ground operations system

AGOSOP Air-Ground Operations Standard Operating Procedures

AHB	attack helicopter battalion
AI	air interdiction
A/L	administrative/logistics
ALO	air liason officer
AM	amplitude modification
ANGLICO	air and naval gunfire liaison company
AO	area of operation
AOE	authorized organization equipment
AP	antipersonnel (mine)
APC	armored personnel carrier
APERS	antipersonnel (animation)

APT	armor-piercing incendiary
ARNG	Army National Guard
ARSOA	Army special operations aviation
ARSOF	Army special operations forces
ARTEP	Army Training and Evaluation Program
ASOC	air support operations center
ASP	ammunition Supply point
AT	antitank
ATGM	antitank guided missile
ATHS	automatic target hand-off system
ATKHB	attack helicopter battalion

ATP ammunition transfer point

AVLJJ	armored vehicle launched bridge
AVUM	aviation unit maintenance
AXP	ambulance exchange point
BAT	battlefield air interdiction
RAS	battalion aid station
B-BAG	battle bags
BDA	battle damage assessment
BDAR	battle damage assessment and repair
BRI	brigade
BDO	battle dress overgarment
BFV	bradley fighting vehicle
BHL	battle handover line
BII	basic issue items
BMNT	beginning morning nautical twilight
BMD	battalion maintenance officer
bn	battalion
Bn/TF	battalion/task force
BP	battle position
BSA	brigade support area

C2	command and control
C2 node	command and control junction

C3 command, control and communication

CA counterair

CAA	combined arms Army
CAB	combat aviation brigade
CAM	chemical agent monitor
CARC	chemical agent resistant coating

CAS	close air support
CBR.	chemical, biological, radiological
CBU	cluster bomb unit
CM	commander
CE	chemical energy; communication-electronics
CEB	clothing exchange and bath
CESO	communications electronics signal officer
CEV	combat engineer vehicle
CFA	covering force area
CFL	coordinated fire line
CFV	cavalry fighting vehicle
CH	cargo helicopter
CLAMMS	cleared lane mechanical making system
cm	centimeter(s)
CMT	company maintenance team
COA	change of assignment course of action
COLT	combat observation lasing team
COMMZ	communications zone
COMSEC	communications security
CONUS	continental United States
COP	command and observation post
COSCOM	corps support command
CP	checkpoint: command post
CPX	command post exercise
CRP	combat reconnaissance patrol
CRT	combat ray tube
CS	combat support
CSB	corps support battalion
CSM	command sergeant major
CSR	controlled supply rate
CSS	combat service support
CSSE	combat service support element
CTA	common table of allowances
CW	continuous wave
D-day	deployment day
DA	Department of the Army
DAG	divisional artillery group
DAO	division ammunition officer
DBDU	desert battle dress uniform
DD	Department of Defense
DEW	director! energy wanton

DISCOM	division support command
DKIE	decontamination kit individual equipment
DLIC	detachment left in conflict
DMD	digital message device
DMMC	division material management center
DNVT	digital nonsecure voice terminal
DP	decision point
DPICM	dual-purpose improved conventional munition
DS	direct support
DSA	division support area
DST	decision support template
DSVT	digital subscriber voice terminal
DTG	date-time group
DTOC	division tactical operations center
DZ	drop zone
EA	engagement area
EAC	echelons above corps
EC	electronic combat
ECM	electronic counter measures
EEFI	essential elements of friendly information
EEI	essential elements of information
EENT	erd evening nautical twilight
EMP	electromagnetic pulse
EN	enemy (in illustration)
EPW	enemy prisoner of war
ETAC	enlisted terminal attack controller
ETL	effective translational lift
EW	electronic warfare
EXTAL	extra time allowance
1SO	first sergeant
FA	field artillery
FAAO	field artillery air observer
FAC	forward air controller
FAAR.	forward area alerting radar
FARP	forward arming refuel point
FASCAM	family of scatterable mines
FAX	facsimile
FDC	fire direction center
FEB A	forward edge of battle area

FED forward entry devices
FFAR folding fin aircraft rocket
FIST fire support team
FIST-V fire support team vehicle
FKSM fort knox supplemental Material
FLIR forward looking infrared
FLOT forward line of own troops
FM field manual; frequency modulation
FO forward observer
FOOGAS petroleum-based expedient Inflammable material
FPF final protective fire
FRAG-HE fragmentary high-explosive (ammunition)
FRAGO fragmentary order
FRG Federal Republic of Germany
FS fire support
FSB forward support battalion
FSCL Fire support coordination line
FSCOORD Fire support coordinator
FSE fire support element
FSO fire support officer
ft feet

G1 assistant chief of staff (personnel)
G3 assistant chief of staff (operations and plans)
G3-Air air operations and planning officer
G4 assistant chief of staff (logistics)
GA gas agent
Gator (a mine delivered by fixed-wing aircraft)
GBU guided bomb unit
GEMSS ground em placed mine scattering system
GLD ground laser designator
GMRD ground motorized rifle division
GPS global positioning systems
GS general support
GSR ground surveillance radar

HE high explosive
HE-APERS high explosive antipersonnel (ammunition)
HEI high explosive incendiary
HEMTT heavy expanded mobile tactical truck

HET	heavy equipment transport
HF	high frequency
HHC	headquarters and headquarters company
HHT	headquarters and headquarters troop
HIMAD	hi-to-medium-altitude air defense
HMMWV	high mobility multipurpose wheeled vehicle headquarters
hq	
HTO	high technology observer
ICM	improved conventional munitions
ID	identification
IDS	infrared discrimination system
IEW	intelligence and electronic warfare
IEWSE	intelligence and electronic warfare support element
IFF	identification, friend or foe
IFFN	identification, friend, foe, or neutral
INTSUM	intelligence summary
IP	initial point
IPB	intelligence preparation of the battlefield
IR	infrared
ITV	improved TOW vehicle
I2	I squared (image intensification)
JAAT	joint air attack team
JTIDS	joint tactical information distribution system
KE	kinetic energy
KIA	killed in action
km	kilometer(s)
kph	kilometer(s) per hour [as a unit of measure only)
kmih	kilometers in the hour
kmph	kilometer(s) per hour (as a unit of measure that indicates motion)
KY	Kentucky
LANTIRN	low-altitude navigation and targeting infrared for night
LAPES	low-altitude parachute extraction system
LBE	load-bearing equipment
LC	line of contact
LCC	land component commander
LCSS	lightweight camouflage screen system
LD	line of departure; low drag
LD/LC	line of departure is line of contact

LGB	laser guided bomb
LO	liaison officer
LQA	light observation aircraft limit of advance
LOC	line(s) of communication
LOG	logistics
LOGPAC	logistics package
LRF	laser range finder
LRP	logistics release point
MAC	military air command
MAGTF	Marine air around task force
medic(s)	medical person: medical personnel
MEF	Marine expeditionary force
MEMO	mission-essential maintenance only
METT-T	mission, enemy, terrain, troops, and time available
MEU	Marine expeditionary unit
MHE	material-handling equipment
Ml	military intelligence
MIC	mid-intensity conflict
MICLIC	mine clearing line charge
mib	miles in the boar
MLC	main line of communication
MLRS	multiple-launch rocket system
mm	millimeter(s)
MMS	multi mission ship
MOPP	mission oriented protection posture
MORT	mortar (in illustration)
MOS	military occupational specialty
MOUT	military operations on urbanized Terrain
MP	military police
MPF	maritime pre- positioning force
mph	miles per hour
MPP	mobile pre-positioning
MPSRON-1	maritime pre-positioning squadron one
MPSRON-2	maritime pre- positioning squadron two
MPSRON-3	maritime pre- positioning squadron three
MPT	mobile pay team
MR	motorized rifle
MRB	motorized rifle battalion
MRC	motorized rifle company

MRD	motorized rifle division
MRE	meals, ready to eat
MRL	multiple rocket launcher
MRP	motorized rifle platoon
MRR.	motorized rifle regiment
MRS	motorized rifle squad
MRS	muzzle reference system
MSB	main support battalion
MSE	mobile subscriber equipment
MSR	main supply route.
MSRT	mobile subscriber radiotelephone terminal
MST	maintenance support team
MTF'	medical treatment facilities
MTLR	moving target locating radar
MTOE	modification table of organization and equipment
MTP	mission training plan
MULE	modular universal laser equipment
NA1	named areas of interest
NATO	North Atlantic Treaty Organization
NBC	nuclear, biological, chemical
NBCWRS	NBC warning and reporting system
NCO	noncommissioned office
NCOIC	non commissioned officer in charge
NCS	net control station
NET	new equipment training
NGLO	naval gunfire liaison officer
NLT	no later than
NOE	night observation device
NOR	nap of the earth
NVD	night-vision device
NVG	night vision goggles
NZO	standing barrier fire
OCOKA	observation and fields of fire, cover and concealment, obstacles, key terrain, avenues of approach
Ol	operations and intelligence
OIC	officer in charge

OIR	other intelligence requirements
OMG	operational maneuver group
OP	observation post
OPCON	operational control
OPLAN	operational plan
OPORD	operation order
OPSEC	operations security
ORP	objective rally point
P&A	personnel and administration
PAC	Personnel and Administrative Center
PADS	position and azimuth determining system
PDS	personnel daily summary reports
PERINTREP	periodic intelligence report
PEWS	platoon early warning system
PIR	priority intelligence requirement
PL	phase line(s)
PHI (	a hybrid PLRS)
PLL	prescribed land list
PLRS	position location reporting system
plt	platoon
POL	petroleum, oils, and lubricants
PMCS	preventive maintenance checks and services
POMCUS	prepositioning of materiel configured to unit sets
PP	passage point(s)
PRF	pulse repetition frequency
PSG	platoon sergeant
PSNC0	personnel services noncommissioned officer
PSO	personnel services officer
PSP	pierced-steel planking
PSS	personnel service support
PST	pass time
PWRS	prepositioned war reserve stock
PX	post exchange
PZ	pick up zone
PZO	rolling harrier fine
QSTAG	Quadripartite Standardization Agreement

RAG	**regimental artillery group**
RAOC	**rear area operations center**
RAP	**rocket-assisted projectile**
RAU	**radio access units**
EG	**repair and evacuation group**
REMS	**remotely employed sensor**
RETRANS	**relay station; retransmit**
RFL	**restrictive fire line**
ROE	**rules of engagement**
ROWPU	**reverse osmosis water purification unit**
RP	**release point**
RPG	**rocket-propelled grenade**
R&S	**reconnaissance and surveillance**
RSR	**required supply rate**
RT	**right; route**
RX	**repairable, exchange**
2IC	**second in command**
S&S	**supply and service**
S1	**adjutant**
S2	**intelligence officer**
S3	**operation and training officer**
S3-Air	**assistant battalion S3 (air operations)**
S4	**supply officer**
S AFAD	**small arms far air defense**
SAM	**surface-to-air missile**
SCT	**scout (in illustration)**
SGM	**sergeant major**
SITREP	**situation report**
SLGR	**small lightweight GPS receiver**
SME	**subject matter expert**
SOI	**signal operation instructions**
SOP	**standing operating procedure**
SP	**shut point**
SPG	**self-propelled grenade**
sq km	**square kilometer(s)**
S&S	**supply and service**
SSB	**single side hand**
STANAG	**Standardisation Agreement**
STB	**supertropical bleach**

TAACOM	Theater Army Air Defense Command
TAC-A	tactical air coordinator-airborne
TACCP	tactical command post
TACCS	Tactical Army Combat Service Support (CSS) Computer System
TACFIRB	tactical line direction system
TACP	tactical air control party
TAI	target areas of interest
TAMMS	The Army Maintenance Management System
TAR	tactical air reconnaissance
TSC	training support center
TC	tank commander
TCF	tactical combat force
TCP	tactical computer processor
TCT	tactical computer terminal
TD	tank division
TDIS	time-distance
TEWT	tactical exercise without troops
TF	task force
TIRS	terrain index reference system
TTS	thermal imaging system
TOC	tactical operations center
TOE	table(s) of organization and equipment
TOT	Time on target
TOW	tube-launched, optically tracked, wire-guided missile
TPU	tank pump unit
TR	tank regiment
TRP	target, reference points
TSOP	tactical SOP

UAV	unmanned aerial vehicle.
UH	Utility helicopter

UHF	ultra high frequency
UMCP	unit maintenance collection point
UMT	unit maintenance team; unit ministry team
US	United States (of America)
USAF	United States Air Force
USMC	United States Marine Corps
USN	United States Navy
VFMED	variable format message entry device
VHF	very high frequency
VINSON	encryption device
vpkm	vehicle per kilometer
VTR	vehicle/tank retriever
WCS	weapons control status
WESS	weapons effect signature simulator
WIA	wounded in action
WO	warning order
WP	white phosphorus
WSM	weapon system manager
WSRO	weapons systems replacement operations
WTR	wide temperature range
XO	executive officer

RDL Table of Document Download
Homepage Contents Information Instructions

REFERENCES

DEPARTMENT OF DEFENSE FORMS (DD Form)

DD Form 565 Statement of Recognition of Deceased. August 1984.

DEPARTMENT OF THE ARMY FORMS (DA Form)

DA Form 581 Request for Issue and Turn-in of Ammunition. August 1989.

DA Form 1155 Witness Statement of Individual. June 1966.

DA Form 1156 Casualty Feeder Report. June 1966.

DA Form 2404 Equipment Inspection and Maintenance Worksheet. April 1979.

DA Form 2765 Request for Issue or Turn-in. April 1966.

DA Form 2765-1 Request for Issue or Turn-in. April 1966.

DA Form 5368-R Quick Fire Plan. December 1984.

FIELD MANUALS (FM)

FM 3-3 NBC Contamination Avoidance. May 1987.

FM 3-4 NBC Protection. October 1985.

FM 3-5 NBC Decontamination. June 1985.

FM 3-50 Smoke Operations. December 1990.

FM 3-100 NBC Defense, Chemical Warfare, Smoke, and Flame Operations. May 1991.

FM 5-34 Engineer Field Data. September 1987.

FM 5-100 Engineer Combat Operations. November 1988.

FM 5-101 Mobility. January 1985.

FM 5-102 Countermobility. March 1985.

FM 5-103 Survivability. June 1985.

FM 6-20 Fire Support in the AirLand Battle. May 1988.

FM 6-20-1 Tactics, Techniques, and Procedures for the Field Artillery Cannon Battalion.

November 1990 .

FM 6-20-40 Tactics, Techniques, and Procedures for Fire Support for Brigade Operations (Heavy). January 1990 .

FM 7-7J The Mechanized Infantry Platoon and Squad (Bradley). February 1986 .

FM 7-10 The Infantry Rifle Company. December 1990 .

FM 7-20 The Infantry Battalion. April 1992 .

FM 7-30 Infantry, Airborne, and Air Assault Brigade Operations. April 1981 .

FM 7-70 The Light Infantry Platoon/Squad. September 1986 .

FM 7-90 Tactical Employment of Mortars. June 1985 .

FM 17-15 Tank Platoon. October 1987 .

FM 17-95 Cavalry Operations. September 1991 .

FM 17-98 The Scout Platoon. October 1987 .

FM 17-98-1 Scout Leader's Handbook. September 1990 .

FM 20-3 Camouflage. November 1990 .

FM 20-32 Mine/Countermine Operations. December 1985 .

FM 21-18 Foot Marches. June 1990 .

FM 21-75 Combat Skills of the Soldier. August 1984.

FM 21-76 Survival. March 1986 .

FM 22-100 Military Leadership. July 1990 .

FM 31-12 Army Forces in Amphibious Operations (The Army Landing Forces). March 1961 .

FM 34-1 Intelligence and Electronic Warfare Operations. July 1987 .

FM 34-2 Collection Management. October 1990 .

FM 34-3 Intelligence Analysis. March 1990 .

FM 34-10 Division Intelligence and Electronic Warfare Operations. November 1986 .

FM 34-80 Brigade and Battalion Intelligence and Electronic Warfare Operations. April 1986 .

FM 34-130 Intelligence Preparation of the Battlefield. May 1989 .

FM 43-5 Unit Maintenance Operations. September 1988 .

FM 43-11 Direct Support Maintenance Operations (Nondivisional). September 1991.

FM 43-12 Division Maintenance Operations. November 1989.

FM 44-1 U S Army Air Defense Artillery Employment. May 1983.

FM 44-3 Air Defense Artillery Employment Chaparral/Vulcan/Stinger. June 1984.

FM 44-16 Chaparral/Vulcan/Stinger Platoon Combat Operations. May 1987.

FM 44-100 United States Army Air Defense Operations. November 1988.

FM 63-3J Combat Service Support Operations - Corps. August 1985.

FM 63-20 Forward Support Battalion. February 1990.

FM 71-1 The Tank and Mechanized Infantry Company Team. November 1988.

FM 71-2 The Tank and Mechanized Infantry Battalion Task Force. September 1988.

FM 71-3 The Armored and Mechanized Infantry Brigade. May 1988.

FM 71-100 Division Operations. June 1990.

FM 71-123 Tactics and Techniques for Combined Arms Heavy Forces: Armored Brigade, Battalion/Task Force, and Company/Team. August 1992.

FM 90-2 Battlefield Deception. October 1988.

FM 90-4 Air Assault Operations. March 1987.

FM 90-6 Mountain Operations. June 1980.

FM 90-10 (HTF) Military Operations in Urban Terrain (MOUT) (How to Fight). August 1979.

FM 90-10-1 (HTF) An Infantryman's Guide to Urban Combat (How to Fight). September 1982.

FM 90-13 (HTF) River Crossing Operations (How to Fight). November 1978.

FM 90-13-1 Combined Arms Breaching Operations. February 1991.

FM 90-14 Rear Battle. June 1985.

FM 90-26 Airborne Operations. December 1990.

FM 100-2-1 The Soviet Army: Operations and Tactics. July 1984.

FM 100-2-2 The Soviet Army: Specialized Warfare and Rear Area Support. July 1984.

FM 100-2-3 The Soviet Army: Troops, Organization, and Equipment. June 1991.

FM 100-5 Operations. May 1986.

FM 100-25 Doctrine for Army Special Operations Forces. December 1991.

FM 100-26 The Air-Ground Operations System. March 1973.

FM 100-103 Army Airspace Command and Control in a Combat Zone. October 1987

FM 101-5 Staff Organization and Operations. May 1984.

FM 101-5-1 Operational Terms and Symbols. October 1985.

FM 101-10-1/2 Staff Officers Field Manual-Organizational, Technical, and Logistical Data, Planning Factors (Volume 2). July 1990.

FLEET MARINE FORCE MANUALS (FMFM)

FMFM 1-5 Maritime Pre-positioning Force (MPF) Operations. Coordinating Draft.

FMFM 4-1 Combat Service Support. September 1981.

FMFM 5-1 Organization and Function of Marine Aviation. October 1991.

FMFM 5-35 Assault Support Helicopter Tactical Manual. March 1992.

FMFM 6-1 Marine Division. March 1978.

FLEET MARINE FORCE REFERENCE PUBLICATIONS (FMFRP)

FMFRP 0-52 Remote Sensing Field Guide, Desert. October 1990.

FMFRP 0-53 A foot in the Desert. October 1990.

FMFRP 0-54 The Persian Gulf Region, A Climatological Study. October 1990,

FMFRP 0-55 Desert Water Supply. November 1990.

FMFRP 0-56 Southwest Asia: Environment and its Relationship to Military Activities. November 1990.

FMFRP 0-57 A Study of Windborne Sand and Dust in Desert Areas. October 1990.

FMFRP 0-58 Problems in Desert Warfare. October 1990.

FMFRP 0-59 The Environment and its Effects on Materiel, Personnel, and Operations with Special Emphasis on the Middle East. October 1990.

FMFRP 0-60 General Design and Construction Criteria for Kuwait.

FMFRP 0-63 Troop Construction in the Middle East. December 1990.

FMFRP 2-12 MAGTF: A Global Capability. April 1991.

GRAPHIC TRAINING AIDS

GTA 3-6-5 NBC Warning and Reporting System.

STANDARDIZATION AGREEMENTS (STANAG)

STANAG 2025 Basic Military Road Traffic Regulations.

STANAG 2050 Statistical Classification of Diseases, Injuries, and Causes of Death.

STANAG 2128 Medical and Dental Supply Procedures.

TRAINING CIRCULARS (TC)

TC 90-1 Military Operations on Urbanized Terrain Training. June 1986.

TECHNICAL MANUALS

TM 3-6665-312-12&P Operator's and Organizational Maintenance Manual Including Repair Parts and Special Tools List for M1A1 Automatic Chemical Agent Alarm. March 1985.

COMMAND PUBLICATIONS

Command publications cannot be obtained through Armywide resupply channels. Determine availability by contacting the installation shown.

FKSM 71-2-8 Battalion Task Force Combat Semite Support Doctrinal Sketch. May 1986.

FKSM 71-8 Close Combat Heavy (CCH) Maneuver Organizations. February 1989.

FKSM 71-10 Movement Coordination Exercise (MCX). August 1989.

ST 17-17 Combined Arms Heavy Battalion/Risk Force SOP. May 1990.

ST 100-3 Battle Book.

For information on the preceding references, contact:

Commander, USAARMC
ATTN: ATZK-IMO-RA
Fort Knox, KY 40121-5000

SH 7-176 Infantry Reference Data Book

For information on the preceding reference, contact:

Commandant, USAIS
ATTN: ATSH-AT
Fort Benning, GA 31905-5000

ST 100-9 The Command Estimate

For information on the preceding reference, contact:

Commander, USACGSC
ATTN: ATZL-SWN-L
Fort Leavenworth, KS 66027-6900

Made in the USA
Middletown, DE
05 November 2026

36611301R00176